AF379914

Michael Granitzer, Mathias Lux and Marc Spaniol (Eds.)

Multimedia Semantics - The Role of Metadata

Studies in Computational Intelligence, Volume 101

Editor-in-chief
Prof. Janusz Kacprzyk
Systems Research Institute
Polish Academy of Sciences
ul. Newelska 6
01-447 Warsaw
Poland
E-mail: kacprzyk@ibspan.waw.pl

Michael Granitzer
Mathias Lux
Marc Spaniol
(Eds.)

Multimedia Semantics - The Role of Metadata

With 73 Figures and 20 Tables

 Springer

Michael Granitzer
Know-Center Graz
Inffeldgasse 21a
A-8010 Graz
Austria
email: mgrani@know-center.at

Mathias Lux
Institut für Informationstechnologie
Universität Klagenfurt
Universitätsstr. 65-67
A-9020 Klagenfurt
Austria
email: mathias.lux@uni-klu.ac.at

Marc Spaniol
Lehrstuhl Informatik V
RWTH Aachen
Ahornstr. 55
D-52056 Aachen
Germany
email: spaniol@dbis.rwth-aachen.de

ISBN 978-3-540-77472-3   e-ISBN 978-3-540-77473-0

Studies in Computational Intelligence ISSN 1860-949X

Library of Congress Control Number: 2008922058

Cover design: Deblik, Berlin, Germany

Printed on acid-free paper

9 8 7 6 5 4 3 2 1

springer.com

Preface

Looking back in history, it is valid to say that metadata in multimedia information is a very old topic. Considering multimedia as information consisting of different media types like drawings and characters, one could argue that the ancient Egypts papyrus is the first known class of multimedia content made and managed by humans. Collecting those multimedia objects in libraries, especially in the famous Bibliotheca Alexandria, Egypts where capable of collecting half a million papyrus book scrolls - a well filled multimedia database nowadays. To maintain and organise such an amount of objects, a clever metadata scheme must have been in use.

Nowadays, in the web and web 2.0 era, a simple search for apples on Flickr results in approximately 500,000 images[1] - a Bibliotheca Alexandria for apples only. In contrast to our ancestors, we do not have human librarians maintaining one single classification or metadata scheme for the worlds known knowledge. Since such an approach would be infeasible or at least too expensive for the information maintained in our world today, we rely on more decentralised approaches based on digital technology. Information is no longer condensed in one single library, but distributed over the whole world. While technical accessibility increases, the labour intensiveness of finding suitable information grows with every piece of information added.

Seeing all currently existing databases as well as the World Wide Web as todays Bibliotheca Alexandria, we are again in charge to develop metadata and classification schemes to maintain this library. This is especially true for multimedia information, since content based access methods available today provide us with somehow usable, but far from perfect access and managing capabilities. Multimedia Metadata - structured linguistic data about multimedia artifacts is required in order to keep multimedia content manage- and accessible for the average user.

However, the quality of assigned metadata depends on various factors, especially on the semantics of the underlying scheme and its accurate usage.

---

[1] http://www.flickr.com/search/?q=apple&w=all, last accessed October 2007

A lot of such metadata schemes exist today, focusing on adding semantics to multimedia items for different domains or different application scenarios. Unfortunately, this variety of standards results in ambiguities, conversion and application problems in real world scenarios, making the laborious acquired metadata nearly useless. In addition, they contribute to the gap in the usage of multimedia standards between research and industry, slowing down essential progress in this domain since research lack of practical data while industry can not easily apply research results to improve their services.

To address those issues, the Multimedia Metadata Community[2] has been founded by research organizations and industry from Austria, France and Germany in 2005. The Multimedia Metadata Community aims at bringing together experts from research and industry in the area of multimedia metadata interoperability for collaborative working environments. By establishing a community of professionals it is intended to bridge the gap between an academic research and an industrial scale development of innovative products for natural collaboration. Our collaboration in this community not only resulted in this book at hand, but also in a series of seven workshops up until now conducted by different members of the community at different locations. Fruitful discussions on various topics led to this book, which presents a summary of topics addressed by our community.

The first part of this book discusses *Fundamentals of Multimedia Metadata Standards* and outlines multimedia metadata formats and their application. Contributions are well balanced between MPEG and non-MPEG based metadata formats. Applications illustrate different usage scenarios for metadata standards (e.g. broadcasting, adaptation and retrieval).

The contribution from Michael Ransburg, Christian Timmerer and Hermann Hellwagner titled "Dynamic and Distributed Multimedia Content Adaptation based on the MPEG-21 Multimedia Framework" introduces the concept of streaming instructions. The streaming instructions are used to facilitate the fragmentation of content-related metadata as well as the linkage between media and metadata in the streaming process of those fragments.

Next, Benoît Le Bonhomme, Marius Preda and Françoise Prêteux introduce a multimedia asset management system around MPEG-4 and MPEG-7. Their contribution "From MPEG-4 Scene Representation to MPEG-7 Description" explains the extraction of MPEG-7 contents by bitstream paring of MPEG-4 contents.

Günther Hölbling, Tilmann Rabl and Harald Kosch give an "Overview of Open Standards for Interactive TV (iTV)". They concentrate on the interactivity in traditional TV by presenting different forms of interaction and giving an overview on existing platforms and standards. In addition, they also introduce the underlying technology of these standards.

The role of metadata in the production process of audiovisual media is introduced by Werner Bailer and Peter Schallauer. Their contribution "Meta-

---

[2] http://www.multimedia-metadata.info/

data in the Audiovisual Media Production Process" analyses the strengths and weaknesses of standards in the production process. They point out that it is crucial to define mappings between these standard in order to ensure interoperability and to overcome the individual shortcomings.

The second section covers a topic of increasingly importance not only in the multimedia domain, namely *Multimedia Semantics*. Contributions in part 2 outline different approaches and describe how semantics may emerge either bottom up or can be defined in a top down process. Applications and described frameworks are pinpointing the capabilities of the different approaches.

Ralf Klamma, Yiwei Cao and Marc Spaniol are considering bottom-up approaches for the capturing of multimedia semantics emerging from a phenomenon called social software. Their article "Smart Social Software for Mobile Cross-Media Communities" addresses issues in metadata interoperability in Web 2.0 applications. Hence, they identify key issues required for creating a software architecture that speaks social software Esperanto.

A top-down approach towards image annotation based on a domain-specific ontology is outlined by Arnaud Da Costa, Eric Leclercq, Arnaud Gaudin, Jean Gascuel and Marie-Noelle Terasse article on "Coupling of an ontology with a modelbased description to combine domain knowledge, metadata and image content: a biological image database example" Their contribution emphasizes on lowering the annotation costs applied in a specialized image database.

The contribution of Cezar Pleşca, Vincent Charvillat and Romulus Grigoraş can be seen as an approach towards bridging the gap between bottom-up and top-down world. Their article on "User-aware adaptation by subjective metadata and inferred implicit descriptors" describes user adaptation in highly dynamic contexts such as varying networks, terminals and user profiles. They introduce a decision-making principle based on reinforcement learning and Markov Decision Processes.

The third part on *Multimedia Retrieval* discusses essential aspects on retrieval of multimedia data, which is a crucial topic crucial for any multimedia driven technology. Nevertheless, multimedia retrieval without taking semantics into account yields to unsatisfactory results in many cases. So since covering the complete topic of multimedia retrieval in depth is impossible, this part focuses on providing the basics of multimedia retrieval and outlines evaluation strategies for measuring the goodness of retrieval approaches. Standardization issues for a homogeneous MPEG-7 Query format conclude the chapter.

Horst Eidenberger and Maia Zaharieva discuss "Semantics in Content-Based Multimedia Retrieval" in order to describe how content-based image features work and what enrichment with semantic information means. First they give an overview of what (semantic) content-features are. Then they describe how these features can be extracted and finally illustrate how these features are used.

Evaluation of multimedia retrieval systems is a critical issue for both, real world applications and research. Mathias Lux, Gisela Granitzer and Günter

Beham present in their contribution on "Multimedia Retrieval Evaluation" based on empirical studies to measure the correlation between user perception and content-based multimedia similarity. Their approach can be used to complement and extend current evaluation approaches.

Mario Döllers article "Specification of an MPEG-7 Query Format" describes the need for a common query format that allows the querying of MPEG-7 enabled multimedia repositories. In addition, an overview of the framework components such as session management, service retrieval and its usability is presented.

The final section concludes the book by outlining *Cross-Modal Multimedia Techniques in Knowledge Management*. It outlines techniques for exploiting relationships in such multimodal documents for visualisation and retrieval based on their intrinsic semantics. It summarises approaches on analysing large sets of multimedia data as well as discussion various cross-modal multimedia techniques in the case of knowledge transfer.

Vedran Sabol, Wolfgang Kienreich and Michael Granitzer describe the difficulties inherent in a manual annotation complex semantics, especially in the context of very large data sets. Their article on "Visualization Techniques for Analysis and Exploration of Multimedia Data" therefore introduces and evaluates various automated methods for the extraction of semantic metadata. Even more, they present visualisation techniques that employ the vast processing power of the human visual apparatus to quickly identify complex patterns in large amounts of data.

Christian Gütls contribution "Automatic Extraction, Indexing, Retrieval and Visualization of Multimodal Meeting Recordings for Knowledge Management Activities" addresses general aspects of multimodal information systems and introduces a conceptual architecture for a generalized view about such systems. In the domain of meeting scenarios an overview about relevant research activities in the context of multimodal information systems is given and practical experiences made in a research project are presented.

Acknowledgements

First of all we thank the chapter authors of this book, for their work and excellent cooperation. Since all chapters in this book have been peer reviewed among the authors and by external reviewers. We want to thank all reviewers for their profound and detailed reviews, ensuring the high quality of the contributions presented here. Reviewers have been

Werner Bailer	JOANNEUM RESEARCH, Austria
Susanne Boll	University of Oldenburg, Germany
Laszlo Böszörmenyi	Klagenfurt University, Austria
Vincent Charvillat	ENSEEIHT, France
Arnaud Da Costa	Bourgogne University, France
Mario Döller	Passau University, Germany
Eidenberger Horst	Vienna University of Technology
Markus Fauster	Klagenfurt University, Austria
Gisela Granitzer	Know-Center Graz, Austria
Romulus Grigoraş	ENSEEIHT, France
Christian Gütl	Graz University of Technology, Austria
Ralf Klamma	RWTH Aachen University, Germany
Werner Klieber	Know-Center Graz, Austria
Harald Kosch	Passau University, Germany
Mark Kröll	Know-Center Graz, Austria
Janine Lachner	Klagenfurt University, Austria
Cezar Pleşca	ENSEEIHT, France
Marius Preda	Institut National des Tlcommunications, France
Michael Ransburg	Klagenfurt University, Austria
Vedran Sabol	Know-Center Graz, Austria
Markus Strohmaier	Graz University of Technology, Austria

We specially want to express our thanks to Harald Kosch from the University of Passau and Ralf Klamma, from RWTH Aachen University who made this book possible and gave us the opportunity to edit it. Last but not least we thank Janusz Kacprzyk from the University of Warsaw, the Editor-in-Chief of this Series, who has given us the opportunity to act as Guest Editors of this book.

Klagenfurt, Graz, Aachen
October 2007

Mathias Lux
Michael Granitzer
Marc Spaniol

Contents

Part III Multimedia Retrieval

Part IV Cross-Modal Multimedia Techniques

Part I

Fundamentals of Multimedia Metadata Standards

Dynamic and Distributed Multimedia Content Adaptation based on the MPEG-21 Multimedia Framework*

Michael Ransburg, Christian Timmerer, and Hermann Hellwagner

Klagenfurt University, Universitätsstraße 65–67, A–9020 Klagenfurt
{michael.ransburg,christian.timmerer,hermann.hellwagner}@uni-klu.ac.at

Today, there are many technologies in place to establish an infrastructure for the delivery and consumption of multimedia content. In practice, however, several elements of such an infrastructure are often stand-alone systems and a big picture of how these elements relate to each other or even fit together is not available. Therefore, MPEG-21 aims to provide an open framework for interoperable multimedia delivery and consumption. This requirement for interoperability results in a great diversity of XML-based metadata, which describes the media data on semantic or syntactic levels, in order to make it more accessible to the user. This metadata can be of considerable size, which leads to problems in streaming scenarios. Other than media data, XML metadata has no concept of samples, thus inhibiting streamed (and timed) processing, which is natural for media data. In order to address the challenges and requirements resulting from this situation, the concept of streaming instructions is introduced. These streaming instructions facilitate the fragmentation of content-related metadata, the association of media and metadata fragments with each other, and the synchronized streaming and processing of those fragments. Based on these capabilities, a dynamic and distributed multimedia content adaptation framework can be built.

1 Introduction and Motivation

The information revolution of the last decade has resulted in an impressive increase in the quantity of multimedia content available to an increasing number of different users with different preferences who access the content through a variety of devices and over heterogeneous networks. End devices range from mobile phones to high definition TVs, access networks can be as diverse as

---

* Work partly supported by the European projects DANAE (IST-1-507113) and ENTHRONE (IST-038463)

M. Ransburg et al.: *Dynamic and Distributed Multimedia Content Adaptation based on the MPEG-21 Multimedia Framework*, Studies in Computational Intelligence (SCI) **101**, 3–23 (2008)
www.springerlink.com

GSM and broadband networks, and the various backbone networks are different in bandwidth and quality of service (QoS) support. In addition, users have different content/presentation preferences and intend to consume the content at different locations, times, and under altering circumstances.

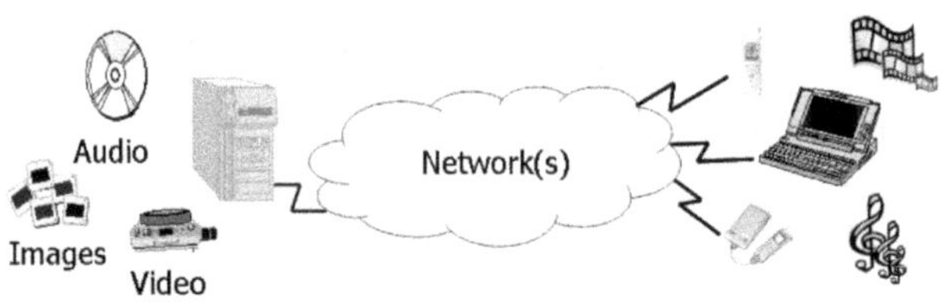

Fig. 1. Concept of UMA [3]

Substantial research and standardization efforts have aimed at supporting *Universal Multimedia Access* (UMA)[34] (Figure 1) which attempts to comply with the scenarios indicated above. The primary goal of UMA is to provide the best QoS or user experience under the given circumstances. The Moving Picture Experts Group (MPEG) supports the concepts provided by UMA by means of normative description tools specified within the MPEG-21 [17][3] standard, specifically in Part 7 which is referred to as Digital Item Adaptation (DIA) [35].

The remainder of this chapter is organized as follows. Section 2 provides a brief overview of the MPEG-21 Multimedia Framework and a more detailed introduction into DIA guided by a use case scenario. The multimedia metadata enabling the processing of media and metadata within streaming environments is described in Section 3. Section 4 provides details on how this kind of metadata can be applied in a DIA scenario. Related work is provided in Section 5 and Section 6 concludes this chapter.

2 The MPEG-21 Multimedia Framework and the Role of Digital Item Adaptation

2.1 The MPEG-21 Multimedia Framework

The aim of the MPEG-21 standard, the so-called Multimedia Framework, is to enable transparent and augmented use of multimedia resources across a wide range of networks, devices, user preferences, and communities, notably for trading of content. As such, MPEG-21 provides the next step in MPEG's standards evolution. The MPEG-21 standard currently comprises 18 parts which can be clustered into six major categories each dealing with different aspects of Digital Items (DIs): declaration and identification, digital rights management, adaptation, processing, systems aspects, and miscellaneous subjects as depicted in Figure 2.

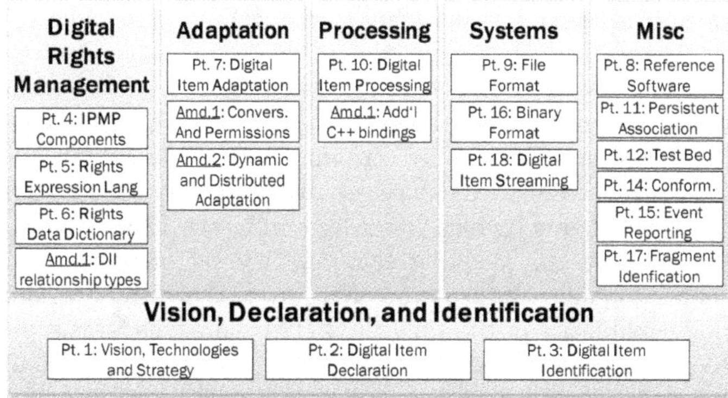

Fig. 2. MPEG-21 Building Blocks

A vital and comprehensive part within MPEG-21, specifically created to foster UMA, is the *Digital Item Adaptation (DIA)* specification, which defines normative description tools to assist with the adaptation of DIs. DIA is addressed in more detail below, based on a use case.

2.2 A Use Case

In the following, a multimedia delivery and consumption scenario is introduced by means of which the scope and role of the MPEG-21 DIA standard can be illustrated. The scenario and architecture described here demand dynamic and distributed content adaptation techniques some of which are at the core of the rest of this chapter.

Assume an Internet Service Provider (ISP) provides a new feature which offers customers live streams of events. Two subscribers use this service to watch a Formula 1 car race. Both persons are in the same room and use their High Definition Television (HDTV) screen to watch this program. After some time, one of them has to leave the room. Since she wants to continue watching the race, she picks her tablet PC and instructs it to duplicate the session from the HDTV screen. The Formula 1 program appears on the tablet PC and she can continue watching the race in another room.

The architecture which the ISP relies upon in order to address this scenario, is explained below. The set-top box (STB) in the subscribers' home needs to facilitate content adaptation and therefore has to be context-aware. When the subscribers start watching the Formula 1 race on their HDTV screen, the STB needs to be aware of the terminal capabilities of the HDTV screen, including its display resolution and its media decoding capabilities. If the requested stream has not already been made available to the STB, the STB forwards the request, including the associated terminal capabilities, to the ISP. The ISP uses the terminal capabilities to adapt the media and starts to stream the selected program (i.e., the adapted media stream) to the STB.

The STB forwards this adapted stream to the HDTV screen where it is being displayed.

When, as described above, one of the customers requests to continue watching the selected program on the tablet PC, the device requests the same channel from the STB and provides its terminal capabilities within the request. The STB analyzes the provided device capabilities in order to find out whether the quality of the channel which it receives from the ISP is appropriate, or if it needs to request the media streams in different quality. In fact in this case, the current quality level is too high given the display resolution and the decoding capabilities of the tablet PC. Therefore, the STB replicates the stream which it already receives in best quality from the ISP and adapts this replicated stream according to the terminal capabilities of the tablet PC.

By means of this setup, the ISP saves bandwidth through *application-layer multicast*. The stream replication and possible adaptation processes are being performed at the customers' premises without putting load on the ISP's equipment. Additionally, the architecture also enables the customers to transfer or duplicate the session to any device in their vicinity. Furthermore, the adaptation techniques employed in this architecture enable the ISP to provide its content to its subscribers anywhere and anytime.

2.3 The Role of the Digital Item Adaptation Standard

For the above scenario to work, interoperability is required between all of the involved devices:

- The end devices (HDTV screen, PDA, tablet PC) need to report their terminal capabilities in a format which is interpretable by both the STB and the ISP's server.
- The STB needs to be able to interpret and adapt the media content which is sent by the server. Ideally, it should be able to perform these actions in a general manner, independently of the actual media encoders used by the server.

MPEG-21 DIA provides normative description formats which enable interoperability in both cases above. *Device interoperability* (i.e., the first item above) is achieved through a unified description model that covers information about the user characteristics, terminal capabilities, network conditions, and natural environment properties. This context information is generally referred to as the *Usage Environment Description (UED)*.

Coding format independence (i.e., the second item above) is accomplished by means of *Bitstream Syntax Descriptions (BSDs)*, *Adaptation Quality of Service (AQoS)* specifications, and *Universal Constraints Descriptions (UCDs)*. The concept of coding format independent multimedia content adaptation relies on the characteristics of scalable coding formats which enable the generation of a degraded version of an original media bitstream by means of simple remove operations followed by minor update operations, e.g., removal of spatial layers of a video and updates of certain header fields conveying the horizontal and vertical resolution. A BSD is an XML document which describes a (scalable) bitstream enabling its adaptation in a codec agnostic way. Only the high-level bitstream structure is described, i.e., how it is organized in terms of packets, headers, or layers. The level of detail of this description depends on the scalability characteristics of the bitstream and the application requirements. In the course of content adaptation, the BSD of a media bitstream is transformed first, followed by the generation of the adapted bitstream from the original one, guided by the modified BSD. An AQoS description provides means to select optimal parameter settings for media content adaptation engines to satisfy constraints imposed by terminals and/or networks while maximizing QoS. In other words, the parameters for transforming the aforementioned BSD are provided. Finally, UCDs restrict the solution space provided by the AQoS description through limitation and optimization constraints.

For further information on the generic BSD-based adaptation process the reader is kindly referred to [9][15][32][7].

3 Processing and Delivery of Metadata for Multimedia Content Streaming

3.1 Motivation and Scope

The MPEG-21 standard as previously introduced shows that the role of XML-based metadata for describing advanced multimedia content gains ever more importance. One purpose of such metadata is to increase the access to such contents from anywhere and anytime. In the past, two main categories for this kind of metadata have become apparent [11]. The first category of metadata aims to describe the *semantics* of the content, e.g., by means of keywords, violence ratings, or classifications. Metadata standards supporting this category are, e.g., MPEG-7, TV Anytime, and SMPTE [1]. The second category of metadata does not describe the semantics, but rather the *syntax and structure* of the multimedia content. This category, for instance, includes languages for describing the bitstream syntax which in turn yield a wide range of research activities enabling codec-agnostic adaptation engines for scalable contents. Examples for such languages are the Bitstream Syntax Description Language (BSDL) as introduced by MPEG-21 DIA, BFlavor [33], and XFlavor [13]. Note that MPEG-7 also provides means for describing syntactical aspects of multimedia bitstreams [2].

Both categories of metadata (semantic and syntactic descriptions) have in common that they tend to be designed in increasing detail, as this increases the accessibility of the media content. They often describe the content per segment or even per access unit (AU), which are the fundamental units for transport of media streams and are defined as the smallest data entities which are atomic in time, i.e., to which decoding time stamps can be attached.

An example of this tendency is that a single violence rating for an entire movie might exclude many potential consumers if it contains only one or two extremely violent scenes. However, if the violence rating were provided per scene, for instance, the problematic scenes could simply be skipped for viewers who are not supposed to see them. On another vein, providing highly descriptive metadata for scalable multimedia content (i.e., describing spatial, temporal, and fine-grained scalability) would enable the accessibility of the content on as many devices as possible.

As a consequence, this metadata is often of considerable size, which even when applying compression is problematic in streaming scenarios. That is, transferring entire metadata files before the actual transmission of the media data – if possible at all – could lead to a significant startup delay. Additionally, there is no information on how this metadata is synchronized with the corresponding media data, which is necessary for streamed (i.e., piece-wise) processing thereof. The concept of piece-wise (and timed) processing is natural for media data. For example, a video consists of a series of independent pictures which are typically taken by a camera. These independent pictures are then encoded, typically exploiting the redundancies between these pictures. The resulting AUs can depend on each other (e.g., in the case of bidirectional encoded pictures) but are still separate samples of data. Although the characteristics of content-related metadata are very similar to those of timed multimedia content, no concept of "samples" exists for this metadata today.

In the following, we introduce the concept of "samples" for metadata by employing streaming instructions for XML-based metadata. Furthermore, streaming instructions for the multimedia content are proposed as well that allow synchronized processing of both media and metadata. The *XML streaming instructions* specify the fragmentation of the content-related metadata into meaningful fragments and their timing. These fragments are referred to as process units (PUs), which introduce the concept of "samples" known from audiovisual content to content-related metadata. The *media streaming instructions* are used to locate AUs in the bitstream and to time them properly. Both types of streaming instructions enable time-synchronized, piece-wise (i.e., streamed) processing and delivery of media data and its related metadata. Furthermore, the fragmentation mechanism helps to overcome the startup delay introduced by the size of the metadata. Another, less obvious benefit is that the streaming instructions enable to extend the existing BSD-based media adaptation approach to dynamic and distributed use cases as the one described in Section 2.2. This extension will be addressed in Section 4.

3.2 Streaming Instructions

In the following, we first introduce the basic requirements which we identified for the streaming of metadata and related media data:

- The streaming instructions need to describe how metadata and/or associated media data should be fragmented into PUs (for metadata) and AUs (for media data) respectively, for processing and/or delivery.
- A PU has to be well-formed (w.r.t. an XML schema) and needs to be able to be consumed and processed as such by a terminal (i.e., no other fragments are needed to consume and process it).
- The streaming instructions shall enable to assign a timestamp to a PU and/or an AU indicating the point in time when the fragment shall be available to a terminal for consumption.
- The streaming instructions need to provide mechanisms which allow a user to join a streaming session that is in progress. This means that one needs to be able to signal when a PU and/or AU shall be packaged in such a way that random access into the stream is enabled.
- It shall be possible to apply the streaming instructions without modifying the original XML document as there may be use cases where it is not possible or feasible to modify the multimedia content and its metadata, e.g., due to digital rights management issues.
- A streaming instruction's processor shall work in a memory and runtime efficient way.

Consequently, we introduce three different mechanisms to respond to the requirements described above:

1. The XML streaming instructions describe how XML documents shall be fragmented and timed.
2. The media streaming instructions localize AUs in the bitstream and provide related time information.
3. Finally, the properties style sheet provides means to describe all of the above properties in a separate document, rather than directly in the metadata. Due to space restrictions, this mechanism is not dealt with in this chapter; further details are given in [31][22].

The XML and media streaming instructions are defined as properties. The properties are abstract in the sense that they do not appear in the XML document, but augment the element information item in the document infoset [5]. They can be assigned to the metadata by using XML attributes and/or by the properties style sheet. Additionally, an inheritance mechanism is defined for some of these properties: the value of the property is then inherited by all descendant elements until the property is defined with a different value which then supersedes the inherited value, and is itself inherited by the descendants. Lastly, a default value is specified for each property.

In the sequel, we will introduce the mechanisms listed above separately and then combine them as they are applied to the scenario in Section 4.

XML Streaming Instructions

The XML streaming instructions provide the information required for streaming an XML document by the composition and timing of PUs. The XML streaming instructions allow firstly to identify PUs in an XML document and secondly to assign time information to them. A PU is a set of connected XML elements. It is specified by one element named anchor element and by a PU mode indicating how other connected elements are aggregated to this anchor to compose the PU. Depending on the mode, the anchor element is not necessarily the root of the PU. Anchor elements are ordered according to the navigation path of the XML document. PUs may overlap, i.e. some elements (including anchor elements) may belong to several PUs. Additionally, the content provider may require that a given PU be encoded as a random access point, i.e., that the encoded PU (the AU) does not require any other AUs to be decoded.

Figure 3 illustrates how an XML document is fragmented and timed using the XML streaming instructions. The fragmenter uses as input the XML document to be streamed and a set of XML streaming instructions properties provided either internally (as XML attributes within the XMLSI namespace) and/or externally (with a properties style sheet as specified in [31][22]). The output of the fragmenter is a set of timed PUs.

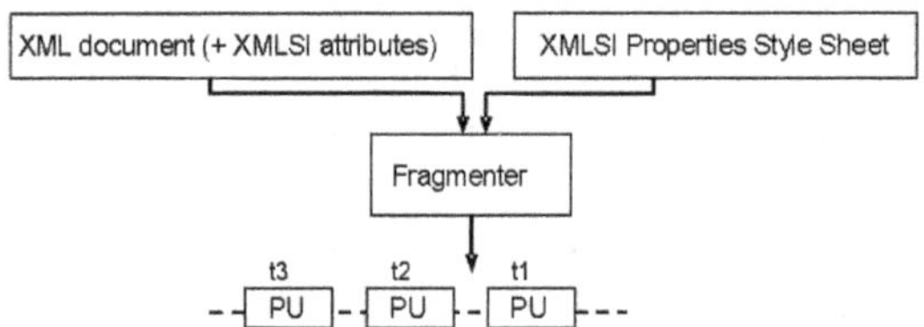

Fig. 3. Processing related to XML Streaming Instructions

The fragmenter parses the XML document in a depth-first order. XML streaming instructions properties are computed as explained below. An element with the *anchorElement* property set to true indicates an anchor element and a new PU. The PU then comprises connected elements according to the *puMode* property of the anchor element.

In the following, the XML streaming instructions properties are specified for:

- Fragmenting an XML document into PUs.
- Indicating which PUs shall be encoded as random access points.
- Assigning time information (i.e., processing time stamp) to these PUs.

The *puMode* property specifies how elements are aggregated to the anchor element (identified by the *anchorElement* property) to compose a PU. Figure

4 gives an overview of the different *puModes*, which were derived by analyzing various types of metadata (as introduced above) and their applications (see Section 4 for a detailed description of an example application). The objective was to constrain ourselves to as few *puModes* as possible, while still supporting all sensible applications, in order to enable an efficient implementation. The semantics of the different *puModes* are defined in Table 1 given that the white node in Figure 4 contains an *anchorElement* property which is set to true.

Table 1. Semantics of Different *puModes*

Name	Semantics
self	The PU contains only the anchor element.
ancestors	The PU contains the anchor element and its ancestor's stack, i.e., all its ancestor elements.
descendants	The PU contains the anchor element and its descendant elements.
ancestorsDescendants	The PU contains the anchor element, its ancestor stack, and its descendant elements.
preceding	The PU contains the anchor element, its descendant elements, its parent element, and all the preceding-sibling elements of the elements part of its ancestor stack and their descendants.
precedingSiblings	The PU contains the anchor element, its descendant elements, its parent elements, and all the preceding-sibling elements (and their descendants) of the element part of its ancestor stack.
sequential	The PU contains the anchor element, its ancestors stack and all the subsequent elements (descendants, siblings and their ancestors) until a next element is flagged as an anchor element.

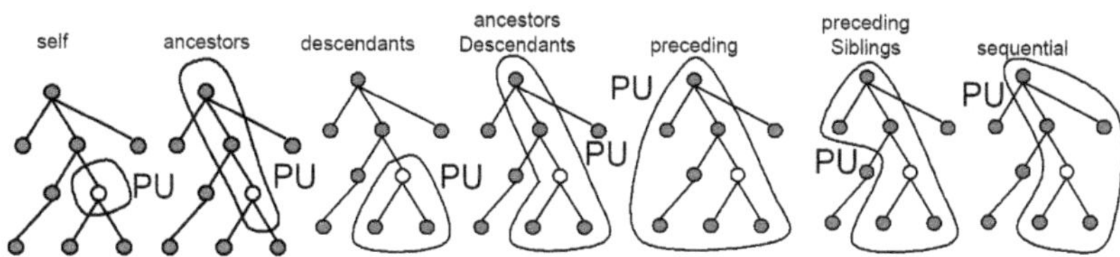

Fig. 4. Examples of Different *puModes*

The *encodeAsRAP* property is used to signal that the PU should be encoded as a random access point in order to enable random access into an XML stream. The *timeScale* property provides the number of ticks per second. The *ptsDelta* property specifies the interval in time ticks after the preceding anchor element. Alternatively, the *pts* property specifies the absolute time of the anchor element as the number of ticks since the origin. The timing can not only be specified in ticks: the *absTime* property specifies the absolute time of the anchor element. Its syntax and semantics are specified according to the time scheme used (*absTimeScheme* property), e.g., NPT, SMPTE or UTC.

Media Streaming Instructions

The media streaming instructions specify two sets of properties for annotating an XML document. The first set indicates the AUs and their location in the described bitstream, the random access points, and the subdivision into AU parts. The second set provides the AU time stamps.

Figure 5 illustrates how AUs in a bitstream are located and timed using the media streaming instructions. The fragmenter uses as input the bitstream to be streamed and a set of media streaming instructions provided either internally (as attributes) and/or externally (with a properties style sheet). The output of the fragmenter is a set of timed AUs.

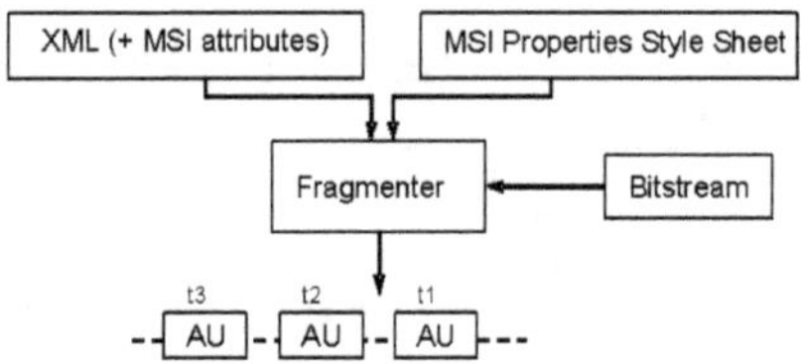

Fig. 5. Processing related to Media Streaming Instructions

The fragmenter parses the XML document in a depth-first order. The media streaming instructions properties are computed as specified below. Anchor elements (i.e., elements with the *au* property set to true) are ordered according to the parsing order and so are the corresponding AUs. An anchor element indicates the start of an AU, the extent of which is specified by the *auMode* property.

In the following, the media streaming instructions properties are specified for:

- Locating AUs in the bitstream.
- Indicating which AUs shall be encoded as random access points.
- Assigning time information (i.e., processing time stamp) to these AUs.

The media streaming instructions are tailored to metadata which can linearly describe a bitstream on an AU granularity, such as BSD, gBSD [35], BFlavor [33], XFlavor [13] or MPEG-7 MDS [28]. The start of an AU is indicated by an element with an *au* property set to true. This element is named anchor element. The media streaming instructions indicate the *start* and the *length* of an AU in bits or bytes (depending on the *addressUnit* property). The extent of the AU depends on the value of the *auMode* property of the anchor element as depicted in Figure 6 (the white node indicates an element with the *au* property set to true). In the *sequential* mode, the AU extends until a new element is found with an *au* property set to false or true. If no element is found with an *au* property set to true or false, the AU extends until the end of the bitstream. In the tree mode, the AU is the bitstream segment described by the XML sub-tree below the element flagged with the *au* property set to true. AU parts are defined in a similar way. The start of a new AU part in an AU is indicated by an *auPart* property set to true and the extent is specified by the *auMode* property. In the sequential mode, the AU part extends until a new element has an *auPart* property set to false or true (in the latter case, a new AU part follows immediately), until the end of the AU, or until the end of the media bitstream. In the tree mode, the AU part is the bitstream segment corresponding to the sub-tree below the element flagged by the *auPart* property. The *auPart* property provides a way for indicating AU parts within an AU in a coding format independent way. In this way, a streaming server that is not aware of the format of the streamed media content may nevertheless meet the requirements of a specific RTP payload format, e.g., special fragmentation rules.

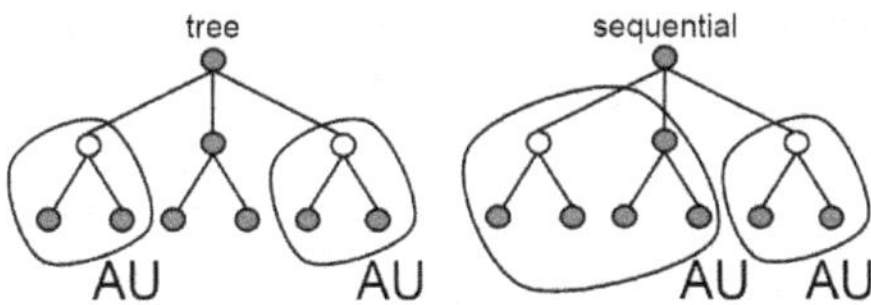

Fig. 6. Examples of Different *auModes*

Other information about AUs is specified by the properties of the anchor element. In particular, the AU is a random access point if the *rap* property of the anchor element is set to true. The *rap* property is inheritable and, thus, it is therefore possible to inherit this property to each AU (i.e., each AU is a RAP) by setting the *rap* property of the XML root element to true. The time information of the AU (CTS and DTS) is also specified by the properties of the anchor element as explained below. The media streaming instructions use an absolute and a relative mode for specifying time information. In absolute mode, the CTS and DTS of an AU are specified independently from other AUs. In relative mode, the CTS and DTS are calculated relatively to the CTS and DTS of the previous AU. Both modes can be used in the same document. For example, an absolute date can be applied to a given AU, and the CTS and DTS of the following AUs are calculated relatively to this AU. In both modes, CTS and DTS conform to a time scale, i.e., they are specified as a number of ticks. The duration of a tick is given by the time scale which indicates the number of ticks per second, which allows for fine granular timing of AUs. The time scale is specified by the *timeScale* property. The two properties *cts* and *dts* define the CTS and DTS of the AU, expressed as an integer number of ticks. They are not inheritable and may be applied to an anchor element for specifying the CTS and DTS of the corresponding AU. Alternatively, two properties named *dtsDelta* and *ctsOffset* allow calculating the DTS and CTS of the AU relatively to the previous AU. The *dtsDelta* property indicates the time interval in ticks between the current AU and the previous one. The *ctsOffset* property indicates the time interval in ticks between the DTS and the CTS of the current AU. Some media codecs do not require a CTS information. In this case, the *cts* and *ctsOffset* properties are not used and may be undefined.

For each anchor element, the properties of the corresponding AU are then calculated as follows:

```
if  isPresent(dts(n))  {   DTS(n)  =  dts(n);  }  else  {
    if  n = 0  {   // i.e.,  first  AU
        DTS(n)  =  0;
    } else  {       DTS(n)  =  ((DTS(n-1)  +  DTS_DELTA(n-1))/TIME_SCALE(n-1))  *
            TIME_SCALE(n)  ;  }
}
if  isPresent(cts(n))  {  CTS(n)  =  cts(n);  }  else  {  CTS(n)  =  DTS(n)  +  ctsOffset;  }
TIME_SCALE(n)  =  timeScale(n);  DTS_DELTA(n)  =  dtsDelta(n);  RAP(n)  =  rap(n);
```

4 Using Streaming Instructions to Enable Dynamic and Distributed Adaptation

MPEG-21 BSD-based adaptation [35] represents a codec-agnostic adaptation approach utilizing XML-based BSDs and exploiting the characteristics of scalable coding formats as described in Section 2.

Due to the fact that the BSD describes the complete bitstream, any adaptation that is performed always impacts the complete bitstream. This is ideal for download and play scenarios, however no piece-wise adaptation to a dynamically changing usage environment is foreseen. Further disadvantages when applying this approach to streaming scenarios include:

- High memory requirements due to the need to parse the complete BSD into memory for the adaptation.
- High startup delay in streaming scenarios, since any adaptation impacts the complete bitstream.
- Slow reaction to dynamically changing usage environment in streaming scenarios, since any adaptation impacts the complete bitstream.

4.1 Approach

This section describes and illustrates how the streaming instructions described above can be used to extend the static MPEG-21 DIA approach towards dynamic and distributed adaptation scenarios. Figure 7 (which represents an extension of the previously introduced BSD-based adaptation system in Section 2) depicts how we integrated the streaming instructions with the BSD-based adaptation approach in an adaptation server in order to enable dynamic and distributed adaptation. The BSD is provided, together with the XML streaming instructions (depicted as two different entities in the figure), to the XML fragmenter. The fragmenter then determines the next PU from the BSD and assigns a time stamp to it, as described in Section 3.2. This PU is then transformed using the XSLT in the same way as a complete BSD would be transformed (as described in Section 2.3). The transformed PU is forwarded to the BSDtoBin processor, which extracts the appropriate media AU and has its time stamp available, thanks to the media streaming instructions. In the next step, the BSDtoBin processor adapts the media AU order to correspond to the transformed PU. The transformed PUs, which are still represented in the text domain, are then encoded into AUs using a proper encoding mechanism. This can for example be a mechanism as basic as a general compression program such as WinZip or gzip. Another possibility would be to use XML-aware compression mechanisms such as XMLPPM [12]. Another way to encode the PUs is to use a specific binary codec for XML such as MPEG's Binary Format for Metadata (BiM) [16]. BiM is a schema-aware encoding mechanism which, if properly configured, removes any redundancy which exists between consecutive PUs. The redundancy, resulting from the requirement that PUs need to be able to be processed independently, is removed and only the new information is encoded into AUs (except in the case when a PU is declared as a RAP). Several studies have been performed on XML compression in the past [4][6][8]. In our own evaluations which also consider streaming support, BiM proved to be the most efficient way to encode PUs [23].

After encoding the PUs into binary AUs, the media and BSD AUs are packetized for transport. In this step, the timing information provided by media and XML streaming instructions is mapped onto the transport layer (RTP in our case), by including it into the packet header. Both the media and BSD AUs are then streamed into the network, where an adaptation proxy could perform additional adaptation steps, or to an end device where the dynamically adapted media is consumed. In this case, the transport of the metadata may be omitted.

Other content-related metadata which does not have fragmentation or timing requirements is not streamed but may be provided using other out-of-band mechanisms. The normative behavior of the MPEG-21 DIA mechanisms is not changed by integrating the streaming instructions.

4.2 Example

In this section we provide example code for the mechanisms described above. Listing 1 shows an MPEG-21 DIA generic BSD (gBSD) which includes media and XML streaming instructions in order to enable dynamic processing of the gBSD and the described media. In this example, each top-level *gBSDUnit* describes an AU of the MPEG-4 Scalable Video Codec[2] [27], including its start and length in bits (as indicated by the *addressUnit* attribute). As can be seen, the BSD already provides attributes for *addressUnit*, *start* and *length*. The fragmenter therefore uses the values in these attributes rather than duplicating them in the corresponding streaming instructions attributes. Within an AU, each *gBSDUnit* describes a single layer of the SVC stream. The layer is identified by the *marker* attribute value, which for the first layer of the second AU states that it is the first temporal layer of the first spatial layer which belongs to the first FGS layer ("T0:S0:F0").

---

[2] All SVC test data was encoded with the SVC reference software, version 4.12.

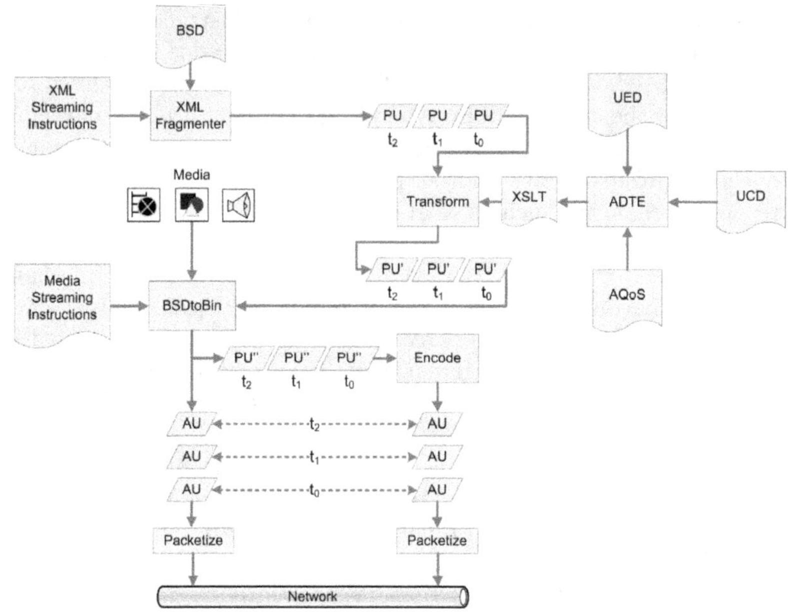

Fig. 7. Dynamic BSD-based Adaptation Approach

Listing 1. Example of gBSD with Streaming Instructions

```
<dia:DIA xmlns:xmlsi="urn:mpeg:mpeg21:200x:01-SI"
    xmlns:msi="urn:mpeg:mpeg21:200x:01-MSI"
    xmlns:dia="urn:mpeg:mpeg21:2003:01-DIA-NS"
    xmlns="urn:mpeg:mpeg21:2003:01-DIA-gBSD-NS"
    xmlns:bs1="urn:mpeg:mpeg21:2003:01-DIA-BSDL1-NS"
    xmlns:xsi="http://www.w3.org/2001/XMLSchema-instance"
    xmlns:xsd="http://www.w3.org/2001/XMLSchema">
<dia:Description msi:timeScale="1000" msi:auMode="tree" xmlsi:timeScale="1000"
    xmlsi:puMode="ancestorsDescendants" xsi:type="gBSDType" addressUnit="bit"
    addressMode="Absolute" bs1:bitstreamURI="cdi_qcif_125_PARLIERsvc_201.raw">
 <gBSDUnit start="0" length="0" msi:dts="0" msi:cts="1280" msi:au="true"
    xmlsi:anchorElement="true" xmlsi:absTimeInt="0" msi:rap="true"
    xmlsi:rap="true">
  <gBSDUnit start="0" length="128" marker="T0:S0:F0" />
  <!-- ... and so on ... -->
  <gBSDUnit start="22288" length="72" marker="T0:S0:F0" />
 </gBSDUnit>
 <gBSDUnit start="22360" length="0" msi:dts="80" msi:cts="1360" msi:au="true"
    xmlsi:anchorElement="true" xmlsi:absTimeInt="80" msi:rap="true"
    xmlsi:rap="true">
  <gBSDUnit start="22360" length="1560" marker="T0:S0:F0" />
  <gBSDUnit start="23920" length="6304" marker="T0:S0:F1" />
  <gBSDUnit start="30224" length="10784" marker="T0:S0:F2" />
  <gBSDUnit start="41008" length="72" marker="T1:S0:F0" />
  <gBSDUnit start="41080" length="1920" marker="T1:S0:F0" />
  <gBSDUnit start="43000" length="552" marker="T1:S0:F1" />
  <gBSDUnit start="43552" length="3048" marker="T1:S0:F2" />
  <!-- ... and so on ... -->
 </gBSDUnit>
 <gBSDUnit start="62992" length="0" msi:dts="1360" msi:cts="2640" msi:au="true"
    xmlsi:anchorElement="true" msi:rap="true" xmlsi:rap="true"
    xmlsi:absTimeInt="1360">
  <gBSDUnit start="62992" length="4456" marker="T0:S0:F0" />
  <!-- ... and so on ... -->
 </gBSDUnit>
 <gBSDUnit start="107616" length="0" msi:dts="2640" msi:cts="3920" msi:au="true"
    xmlsi:anchorElement="true" xmlsi:absTimeInt="2640" msi:rap="true"
    xmlsi:rap="true">
  <gBSDUnit start="107616" length="1200" marker="T0:S0:F0" />
  <!-- ... and so on ... -->
 </gBSDUnit>
 <!-- ... and so on ... -->
</dia:Description>
```

```
</dia:DIA>
```

The streaming instructions are indicated by their respective namespaces (XMLSI and MSI). After declaring the namespaces which belong to the streaming instructions, the *timeScale, auMode* and *puMode* are specified in the Description element. The inheritance of these properties makes sure that they are valid for all *gBSDUnits* which are children of the Description element. In this application, the *ancestorsDescendants puMode* is used, which specifies that any PU consists of the element containing the *anchorElement* attribute and all its ancestors and descendants. The first resulting PU, when applying this fragmentation rule, can be seen in Listing 2. Investigation of these documents shows that each document describes only a small part (in this case an AU) of the media bitstream. However, as we used the *ancestorsDescendants puMode*, the documents correspond to the requirement that a PU has to be well-formed and needs to be able to be consumed as such by a terminal. This allows us to use normative DIA mechanisms without the need to change them. These PUs are then provided to the BSDtoBin processor, which extracts the AUs, as specified by the media streaming instructions and adapts them, as specified by MPEG-21 DIA.

Listing 2. First PU resulting from Processing the gBSD in Listing 1

```
<dia:DIA < !-- NS declarations ommited to save space --> >
 <dia:Description  msi:timeScale="1000"  msi:auMode="tree"  xmlsi:timescale="1000"
     xmlsi:puMode="ancestorsDescendants"  xsi:type="gBSDType"  addressUnit="bit"
     addressMode="Absolute"  bs1:bitstreamURI="cdi_qcif_125_PARLIERsvc_201.raw">
  <gBSDUnit start="0"  length="0"  msi:dts="0"  msi:cts="1280"  msi:au="true"
      xmlsi:anchorElement="true"  xmlsi:absTimeInt="0">
  <gBSDUnit start="0"  length="128"  marker="T0:S0:F0"/>
  <!-- ... and so on ... -->
  </gBSDUnit>
 </dia:Description>
</dia:DIA>
```

4.3 Validation and Performance

In order to validate our work, the system described in this section was implemented in C++ [24], together with the streaming instructions processors, i.e., the media and XML fragmenter. The libxml XMLTextReader interface[3] (an XML Pull Parser) was used for accessing the XML information. The aim of the measurements is to evaluate if our prototype implementation of a dynamic MPEG-21 adaptation node can be utilized in a real-time streaming scenario. To this end, we first measure the performance of the streaming instructions processors and then we evaluate the CPU load and memory utilization of the complete adaptation node (depicted in Figure 9). All tests were performed on a Dell Optiplex GX620 desktop with an Intel Pentium D 2.8 GHz processor and 1024 MB of RAM using Windows XP SP2 as an operating system. Time measurements were performed using the ANSI-C clock method.

Table 2 provides an overview of the test data. Media and the corresponding BSDs for three different media codecs were selected. MPEG-4 BSAC [20] is a

---

[3] libxml, http://xmlsoft.org

Table 2. Characteristics of Test Data

	MPEG-4 BSAC	EZBC	MPEG-4 SVC
Media Size [kB]	12511	450536	538816
Average AU Size [kB]	0.22	197.86	18.59
BSD Size [kB]	196265	144939	123189
Average PU Size [kB]	4.02	63.80	4.90
Number of AUs/PUs	56100	2277	28980
Resolution	N/A	QCIF	QCIF
Frame Rate [fps]	21	12.5	12.5
Length [min]	44.52	48.58	193.20

scalable audio codec, EZBC [14] is a scalable video codec based on wavelets
and MPEG-4 SVC [27] is a scalable video codec based on conventional block
transforms which is currently being standardized in MPEG. The considerable
size differences between the SVC and the EZBC content (both media and
metadata) are due to the fact that the EZBC was encoded with 6 spatial
layers and the SVC was encoded with only a single spatial layer. For our
tests, the BSD is provided in the uncompressed domain and we consider that
each PU describes exactly one AU. We used streaming instructions embedded
into the BSD to specify the fragmentation mechanism for our measurements.
All tests have been repeated 10 times in order to get accurate results.

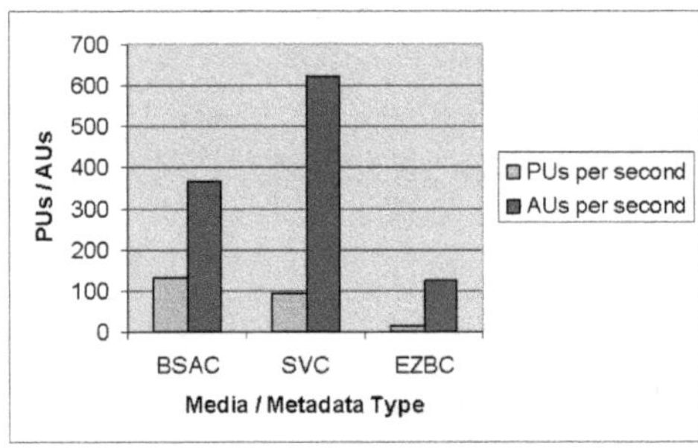

Fig. 8. Streaming Instructions: Performance

For the XML fragmenter the measurements cover: 1) Access to the BSD
from the file system, 2) Parsing the BSD using the libxml XMLTextReader, 3)
Composing PUs, 4) Assigning timing information to the PUs, and 5) Encap-
sulating PUs and their timing into RTP packets. For the media fragmenter
the measurements cover: 1) Access to the BSD from the file system, 2) Access
to the media from the file system, 3) Parsing the BSD using the libxml XML-
TextReader, 4) Extracting AUs, 5) Assigning timing information to the AUs,
and 6) Outputting AUs and their timing to a file. Figure 8 shows the performance of the media and XML fragmenters. Considering that each EZBC AU
describes 16 temporal layers (i.e., frames) and that each SVC AU describes
5 temporal layers, we can conclude that our prototype implementation offers
good real-time performance.

Subsequently, we measured the performance of the complete adaptation node, as depicted in Figure 7.

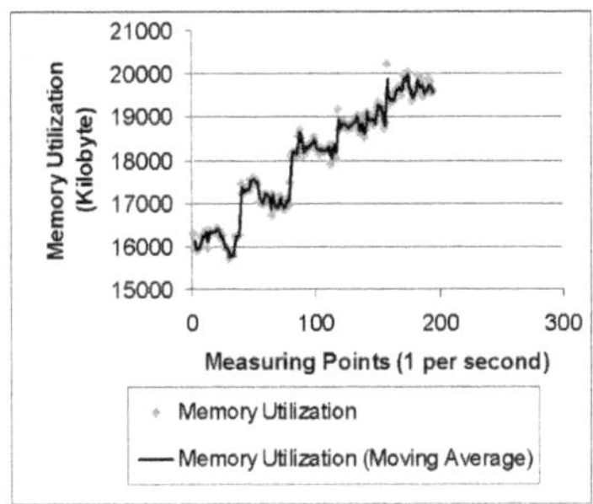
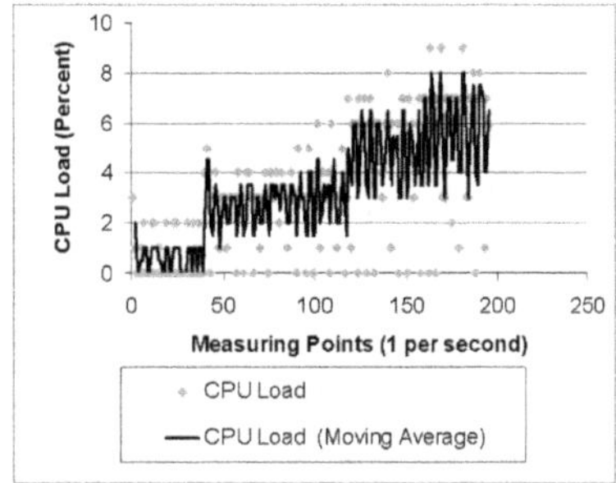

Fig. 9. MPEG-21 based Dynamic DIA Adaptation of 1 to 5 QCIF SVC Streams: Memory Utilization and CPU Load

We measured the memory utilization and CPU load of our adaptation server. To this end, we access a single content (consisting of a media stream and a BSD), fragment it according to the streaming instructions, adapt, packetize and stream it to the player on the end device. We then access another content, and so on, until there are ten streams (five media streams and five BSD streams) being processed and delivered concurrently. Figure 9 shows the results of these tests for the SVC content. There is a single content being processed for the first 40 seconds; then there are two contents until second 80; and so on, with every 40 seconds a new stream being added and processed in parallel. Additional evaluations for other kinds of scalable media have been published in [25]. As can be seen from the measurements, the adaptation node would have supported several more content streams (or contents with a higher bitrate).

5 Related Work

In this section we review related work in the literature that deals with mechanisms enabling streamed processing and transport of multimedia content and related metadata. Multiple mechanisms for specifying the fragmentation and timing of media content are well known, e.g., the sample tables of the ISO Base Media File Format [30]. The difference is that in our approach this information is specified as a part of the metadata. This coupling provides a common way for a user to specify the fragmentation and timing of both media and metadata.

MPEG is currently standardizing so called Multimedia Application Formats, which aim at combining technology from MPEG and other standardization bodies to specify a specific application, e.g., a photo player and a music player [10]. All these applications employ XML metadata and currently either use it only on a track/movie level or they use mechanisms from the ISO Base Media File Format to provide the timing of more dense metadata. However, this requires that the metadata is already fragmented beforehand and that the metadata is therefore no longer available in its original format for non-streamed processing.

Wong et al. [36] define a method for fragmenting an XML document for optimized transport and consumption, preserving the well-formedness of the fragments. However, what is consumed are not the fragments themselves but rather the document resulting from the aggregation of the fragments. Furthermore, the fragmentation is achieved according to the size of the Maximum Transport Unit (MTU) and not based on the semantics of the fragment, i.e., no syntax is provided for a content author to specify which fragments should be consumed at a given time.

Alternatively, MPEG-7 provides an encoding method (Binary Format for Metadata) to progressively deliver and consume XML documents in an efficient way [16]. Therefore, so-called Fragment Update Units (FUUs) provide means for altering the current description tree by adding or removing elements or attributes. However, MPEG-7 only specifies the syntax of FUUs and its decoding, whereas our work concentrates on the composition of XML fragments.

In both cases above, no timing information is provided which enables the synchronized use of the metadata and the corresponding multimedia content.

The Continuous Media Markup Language (CMML) [18] is an XML-based mark-up language for time-continuous data similar to MPEG-7. Together with the Annodex exchange format [19] it allows to interleave time-continuous data with CMML markup in a streamable manner. This approach is specific to CMML whereas in our work we aim to offer a generic solution for time-synchronized, streamed processing and transport for media and related metadata.

The Synchronized Multimedia Integration Language (SMIL) [26] provides a timing and synchronization module which can be used to synchronize the play-out of different media streams. However SMIL is only concerned with media as a whole and therefore no AU location, fragmentation, and timing for metadata are provided.

The Simple API for XML (SAX) is an event-based API which allows streamed processing of XML [29]. It allows to parse an XML document without loading the complete document into memory. This does help to avoid the startup delay for streamed processing. However, legacy applications which rely on DOM would need to be re-implemented. Moreover, no timing or fragmentation information is provided for piece-wise and synchronized processing of media and metadata. However, SAX might further increase the performance of our current implementation where we currently use an XML Pull Parser.

Our concept is close to a mechanism provided by Scalable Vector Graphics (SVG) [21] to indicate how a document should be progressively rendered: the *externalResourcesRequired* attribute added to an element specifies that the document should not be rendered until the sub-tree underneath is completely delivered. This mechanism is specific to SVG. In contrast, our method allows isolating a fragment that can be consumed at a given time, but this fragment does not need to contain the previous one. In particular, it is possible to progressively consume a document without ever the need of loading the full document into memory since only a fragment is consumed at a time.

To the best of our knowledge, the concept of PU and in particular the method we developed for specifying their composition, processing, and their transport in conjunction with media fragments is therefore original.

6 Conclusion

In this chapter, we introduced the MPEG-21 Multimedia Framework which shows how multimedia metadata can contribute to approaching the UMA vision. Consequently we addressed the problem of processing large metadata descriptions in streaming scenarios. To this end, we introduced streaming instructions for fragmenting content-related metadata, associating the media segments and metadata fragments with each other, and streaming and processing them in a synchronized manner. The streaming instructions extend an XML metadata document by providing additional attributes to describe the fragmentation and timing of media data and XML metadata such as to enable their synchronized delivery and processing. In addition, a style sheet approach provides the opportunity to dynamically set such streaming properties without actually modifying the metadata. We evaluated the implemented mechanisms both as stand-alone processors and integrated in a specific application scenario. We showed the usefulness of our work by implementing an adaptation node which uses our mechanisms to extend the static DIA approach to dynamic and distributed usage scenarios.

The streaming instructions have been proposed for inclusion in the MPEG-21 Multimedia Framework and are currently being standardized as an amendment to DIA [31].

References

1. R.S. Atarashi, J. Kishigami, and S. Sugimoto. Metadata and new challenges. In *Symposium on Applications and the Internet Workshop*, January 2003.
2. W. Bailer and P. Schallauer. Detailed audiovisual profile: enabling interoperability between MPEG-7 based systems. In *12th International Multi-Media Modeling Conference*, Beijing, China, January 2006.
3. I. Burnett, R. Koenen, F. Pereira, and R. Van de Walle, editors. *The MPEG-21 Book*. Wiley, 2006.
4. M. Cokus and D. Winkowski. XML Sizing and Compression Study For Military Wireless Data. In *XML Conference & Exposition*, December 2002.

5. J. Cowan and R. Tobin, editors. *XML Information Set (Second Edition)*. World Wide Web Consortium, 2004.

6. S.J. Davis and I. Burnett. Efficient Delivery within the MPEG-21 Framework. In *First International Conference on Automated Production of Cross Media Content for Multi-Channel Distribution*, pages 205–208, Florence, Italy, November 2005.

7. D. De Schrijver, W. De Neve, K. De Wolf, R. De Sutter, and R. Van de Walle. An optimized MPEG-21 BSDL framework for the adaptation of scalable bitstreams. *Journal of Visual Communication and Image Representation*, 18(3):217–239, June 2007.

8. R. De Sutter, S. Lerouge, P. De Neve, C. Timmerer, H. Hellwagner, and R.V. de Walle. Comparison of XML serializations: cost benefits versus complexity. *Multimedia Systems Journal*, 12(2):101–115, 2006.

9. S. Devillers, C. Timmerer, J. Heuer, and H. Hellwagner. Bitstream Syntax Description-Based Adaptation in Streaming and Constrained Environments. *IEEE Transactions on Multimedia*, 7(3):463–470, June 2005.

10. K.P. Diepold and F.W. Chang. MPEG-A: Multimedia Application Formats. *IEEE Multimedia*, 12(4):34–41, October 2005.

11. P. Fox, D. McGuinness, R. Raskin, and K. Sinha. Semantically-Enabled Scientific Data Integration. In *Geoinformatics 2006*, May 2006.

12. S. Harrusi, A. Averbuch, and A. Yehudai. XML Syntax Conscious Compression. In *Data Compression Conference*, pages 402–411, March 2006.

13. D. Hong and A. Eleftheriadis. XFlavor: Bridging Bits and Objects in Media Representation. In *IEEE International Conference on Multimedia and Expo*, Lausanne, Switzerland, August 2002.

14. S.-T. Hsiang and J. W. Woods. Embedded image coding using zeroblocks of subband/wavelet coefficients and context modelling. In *MPEG-4 Workshop and Exhibition at ISCAS 2000*, Geneva, Switzerland, May 2000.

15. A. Hutter, P. Amon, G. Panis, E. Delfosse, M. Ransburg, and H. Hellwagner. Automatic Adaptation of Streaming Multimedia Content in a Dynamic and Distributed Environment. In *International Conference on Image Processing*, Genova, Italy, September 2005.

16. U. Niedermeier, J. Heuer, A. Hutter, W. Stechele, and A. Kaup. An MPEG-7 tool for compression and streaming of XML data. In *IEEE International Conference on Multimedia and Expo*, Lausanne, Switzerland, August 2002.

17. F. Pereira, J. Smith, and A. Vetro. Special Section on MPEG-21. *IEEE Transactions on Multimedia*, 7(3), June 2005.

18. S. Pfeiffer, C. Parker, and A. Pang. The Continuous Media Markup Language. Technical report, Internet Engineering Task Force, March 2004. Internet Draft.

19. S. Pfeiffer, C. Parker, and A. Pang. The Annodex exchange format for time-continuous bitstreams. Technical report, Internet Engineering Task Force, March 2005. Internet Draft.

20. H. Purnhagen. An Overview of MPEG-4 Audio Version 2. In *17th International Conference on High-Quality Audio Coding*, Florence, Italy, September 1999.

21. A. Quint and F. Design. Scalable vector graphics. *IEEE Multimedia*, 10(3):99–102, July 2003.

22. M. Ransburg, S. Devillers, C. Timmerer, and H. Hellwagner. Processing and Delivery of Multimedia Metadata for Multimedia Content Streaming. In *6th Workshop on Multimedia Semantics - The Role of Metadata*, Aachen, Germany, March 2007.

23. M. Ransburg, C. Timmerer, and H. Hellwagner. Transport Mechanisms for Metadata-driven Distributed Multimedia Adaptation. In *First International Conference on Multimedia Access Networks*, pages 25–29, July 2005.

24. M. Ransburg, C. Timmerer, and H. Hellwagner. Dynamic and Distributed Adaptation of Scalable Multimedia Content in a Context-Aware Environment. In *European Symposium on Mobile Media Delivery*, September 2006.

25. M. Ransburg, C. Timmerer, H. Hellwagner, and S. Devillers. Design and evaluation of a metadata-driven adaptation node. In *International Workshop on Image Analysis for Multimedia Interactive Services*, Santorin, Greece, June 2007.

26. L. Rutledge. SMIL 2.0: XML for Web multimedia. *IEEE Internet Computing*, 5(5):78–84, September 2001.

27. H. Schwarz, D. Marpe, and T. Wiegand. Overview of the Scalable H.264/MPEG4-AVC Extension. In *International Conference on Image Processing*, October 2006.

28. T. Sikora. The MPEG-7 visual standard for content description—an overview. *IEEE Transactions on Circuits and Systems for Video Technology*, 11(6):696–702, June 2001.

29. F. Simeoni, D. Lievens, R. Conn, and P. Mangh. Language bindings to XML. *IEEE Internet Computing*, 7(1):19–27, January 2003.

30. D. Singer, editor. *ISO/IEC 14496-12:2005 Part 12: ISO Base Media File Format*. International Organization for Standardization, 2005.

31. C. Timmerer, S. Devillers, and M. Ransburg, editors. *ISO/IEC 21000-7:2004/FPDAmd 2: Dynamic and Distributed Adaptation*. International Standardization Organization, 2006.

32. C. Timmerer and H. Hellwagner. Interoperable adaptive multimedia communication. *IEEE Multimedia Magazine*, 12(1):74–79, January 2005.

33. D. Van Deursen, W. De Neve, D. De Schrijver, and R. Van de Walle. BFlavor: an optimized XML-based framework for multimedia content customization. In *25th Picture Coding Symposium*, Beijing, China, April 2006.

34. A. Vetro, C. Christopoulos, and T. Ebrahimi. Special Issue on Universal Multimedia Access. *IEEE Signal Processing Magazine*, 20(2), March 2003.

35. A. Vetro and C. Timmerer. Digital Item Adaptation: Overview of Standardization and Research Activities. *IEEE Transactions on Multimedia*, 7(3):418–426, June 2005.

36. E. Y. C. Wong, T. S. Chan, and H. Leong. Semantic-based Approach to Streaming XML Contents using Xstream. In *27th Annual International Computer Software and Applications Conference*, Dallas, TX, USA, November 2003.

From MPEG-4 Scene Representation to MPEG-7 Description

Benoît Le Bonhomme, Marius Preda and Françoise Prêteux

ARTEMIS Department, GET-INT, 9 rue Charles Fourier, 91011 Evry, France, www-artemis.int-evry.fr

This chapter introduces the first online MAMS (Multimedia Asset Management System) synergistically developed around MPEG-4/MPEG-7 standards. We show how this combination solves the main challenges for any MAMS with respect to multimedia compression, representation, description as well as user-interactivity. MPEG-4 can be used as a unique solution for storage and presentation of image, audio, video and 2D/3D graphics. Moreover, the basic MPEG-4 components (*e.g.* the elements of scene graphs) are easily extracted by bitstream parsing and exploited for completing the MPEG-7 content description schema. The solution we provide is open to third party descriptor extractors; with this purpose, a developer API (Application Programming Interface) grants access to decoded media and to the instantiation of the descriptors. The practical validation is achieved by an online MAMS devoted to the indexation of 3D graphics objects corresponding to large and heterogeneous databases.

1 Introduction

Today, producing media as image, audio and video is easier than ever. A proliferation of capturing devices (cameras, smart phones, ...) together with smooth integration of communication between them based on open standards makes that the yesterday's multimedia consumers become easily content producers. Moreover, during the last years, online web-based platforms become standard solutions for sharing and distributing media. Within the multimedia world, this means the shift from the *one*-producer-

B.L. Bonhomme et al.: *From MPEG-4 Scene Representation to MPEG-7 Description*, Studies in Computational Intelligence (SCI) **101**, 25–44 (2008)
www.springerlink.com

to-many-consumers paradigm (traditional in the television world) to a *many-to-many* approach (Internet based-sharing) which is now in progress. The goal of this chapter is to present the main components of an open and standard-based online multimedia sharing system. The developed system, called MyMultimediaWorld.com and referred to as MMW.com, aims at offering a step forward in solving the main bottlenecks of the current systems.

- The first functionality of any MAMS (Multimedia Asset Management System) is to support input media of different types and formats. However, most research efforts were focused on multimedia databases managing a single media type (video, audio, or image). Muvis [16], VideoQ [6], Photobook [24], Visualseek [30], Calif & Emir [19], SCHEMA [21] for image/video/audio or the Princeton database [10] and www.3dvia.com for 3D objects are representative for these efforts. Such systems implement software modules enabling the conversion from any format into an internal one. This internal format is specific to each system and differs for each media type. It has a deep impact on the system performances since it should support data compression, easy decoding and streaming. In MMW.com, the MPEG-4 format is adopted for its capability to support in a unified manner video, still picture, audio, graphics, as well as their combination into a complex scene. To make MMW.com operational, MPEG-4 convertors for a large family of multimedia formats were developed.
- The second objective of any MAMS is to index the content for easy retrieval. Therefore, feature extraction and content description algorithms as well as software modules have to be implemented. Currently, semantic indexation of hybrid commercial databases is mainly achieved by using textual annotations and low level descriptors, as colour/forms for images or motion/time segmentation for videos. For a global overview of multimedia database management and indexation as well as of their implementation rules, readers are invited to refer to [14, 13, 2]. Special data models for video are introduced in [11, 26], where videos are first decomposed in temporal entities (segments, scenes, frames) and features are then attached to each entity. Examples of combinations of low level features and their relationships are also reported in [16, 29]. To overcome these limitations, MMW.com is based on the MPEG-7 schema for describing the generic multimedia content. Additionally, this basic schema is extended and enriched by some descriptions directly extracted from the MPEG-4 scene graph and from some graphics primitives. One major contribution of MMW.com is to provide the users with a software API independent of the media type in order to support any feature ex-

traction module. The major advantage of this functionality is to achieve very easily the integration of specific descriptor extraction algorithms and to enable testing and benchmarking on a common database.

- An efficient MAMS has to set up the service management including user query and community management. Commercial systems available today are based on textual query while academic prototypes rely on query by example [21, 9]. Combinations of automatic temporal segmentation and textual annotation are implemented in [1]. More recently, systems like [19, 26] combine high-level and low-level information. Additionally, MPEG is currently developing the MPEG-7 Query Format [32] for standardising the query/answer formalism and a set of requirements [7] for the interface to the multimedia databases. MMW.com implements the two nowadays most intensively considered approaches for formulating the user request, namely SQL and XML and can be easily adapted for MPEG-7 Query Format.
- Finally, any MAMS relies on a presentation engine. Current systems adopt a player for each type of media, from simple image viewers to more complex 3D graphics players. In MMW.com, the presentation engine is powered by an integrated MPEG-4 player able to unitarily handle image/audio/video/graphics.

The "high-level" architecture of MMW.com, outlining the above-mentioned key aspects, is illustrated in Fig. 1.

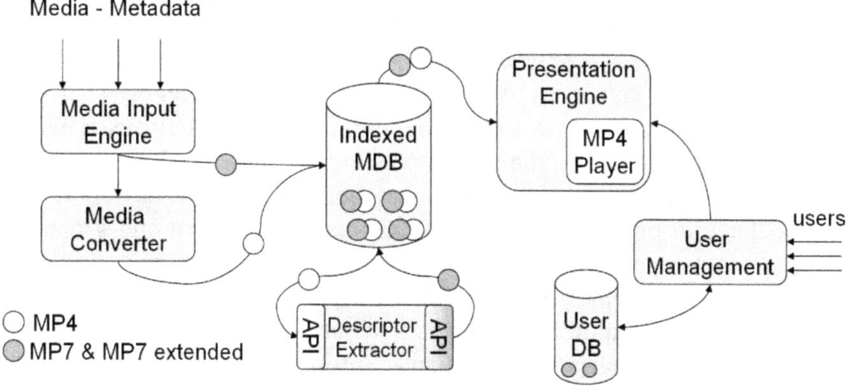

Fig. 1. The "high-level" synopsis of the architecture of MMW.com exhibiting the four contributions introduced: a unique standardised media format, a unique standardised description format and language, an open multimedia database, and a unique rendering engine making the MAMS modular and user-friendly system

The chapter is organised as follows. Section 2 presents an overview of the MPEG-4 tools, the main architecture as well as some details on the standardised tools for the scene graph representation and 3D graphics primitives. Section 3 describes the extended and enriched MPEG-7 schema we developed. Elements directly extracted from the MPEG-4 scene graph or computed by applying specific algorithms are described. The API making it possible content description is presented and its implementation for 3D objects detailed. The last section concludes this chapter and sketches the perspectives for using the proposed system in different application fields.

2 MPEG-4: a rich standard for multimedia representation

MPEG-4 is not only an improved version over its predecessors, MPEG-1 and MPEG-2, with respect to the audio/video compression performances. MPEG-4 is a standard which extends the class of media types by including 2D/3D graphics, text and synthesised audio and allows their composition, enabling the representation of the so-called multimedia scenes. When dealing with synthetic content, the frontier between representation (*i.e.* the set of parameters specifying how to render the content) and the description (*i.e.* the set of parameters defining the key characteristics of the content and serving for indexing and retrieval purposes) becomes fuzzy. So fuzzy that we may find in the literature the term "scene description" for defining how a scene should be rendered. Part 11 of the MPEG-4 standard entitled *"Coding of audio-visual objects - Part 11: Scene description and application engine"*, also known as BIFS (BInary Format for Scenes) [12] provides an example of such a fuzzy boundary. BIFS standardises how the elementary media and graphics primitives are combined, what interaction properties per object are exposed to the user and how the different elements behave in time. For the synthesised content (*i.e.* text and graphics) MPEG-4, as a representation format, specifies how the media should be displayed by the terminal, and implicitly becomes a description format. At the low level, the graphical primitives such as a "rectangle", "circle" and "sphere" include some intrinsic semantic information that can be directly exploited by a search engine. At high level, MPEG-4 content is structured in a scene graph. The relationships between the objects in the scene are intrinsic to the content. A search engine can easily navigate into branches of the scene graph, interrogate properties of the objects (like shape, colour, and trajectory), and retrieve objects based on their properties or sub-graphs based on their configuration.

Under this framework, the way in which the scene is represented becomes a crucial issue. In the last decade, the hyperlink and composition principles of different sources initially developed for the text over Internet have been extended to media scenes, resulting in the development of plenty of scene representation formalisms. Some of them are open standards (like VRML[1], X3D[2], SMIL[3], SVG[4], MPEG-4 BIFS, MPEG-4 XMT, MPEG-4 LaSER, COLLADA[5]) published by ISO, W3C or Khronos; others, like Flash by Adobe or 3ds by Autodesk, accompanying authoring tools, become *de facto* standards.

In the next paragraphs we first recall the multimedia scene graph concept, and review the related existing models and implementations. Then we describe the powerful MPEG-4 scene graph modelling. The spatial and temporal relationships between media and scene elements are defined. For the specific 3D graphics data we show how attributes exposed in the scene graph may be directly exploited to describe the content. Finally, the representation tools standardised by MPEG-4 for 3D graphics objects are revisited for exhibiting in each case which attributes may be used to index the content.

2.1 The Scene Graph concept and its implementations

The scene-graph is the general data structure commonly used by vector-based graphics editing applications and modern computer games. This structure arranges the logical and spatial representation of a multimedia scene and corresponds to a collection of nodes organised as a graph or a tree. A node may have many children but often only a single parent. An operation applied to a parent automatically propagates its effect to all of its children. In many scene graph implementations, associating a geometrical transformation matrix at each level and concatenating such matrices together is an efficient and natural way to perform animation. A common feature is the ability to group related shapes/objects into a compound object which can then be moved, transformed, selected, *etc.* as easily as a single object.

Historically, the scene graph concepts are inherited from the techniques enabling to optimise 3D graphics rendering. In order not to process invisi-

---

[1] Virtual Reality Modelling Language
[2] www.web3d.org
[3] Synchronized Multimedia Integration Language, www.w3.org/AudioVideo/
[4] Scalable Vector Graphics, www.w3.org/Graphics/SVG/
[5] www.collada.org/mediawiki/index.php/Main_Page

ble objects (*i.e.* objects being outside the view frustum) a logic organisation of relations between them is created. Additionally, common properties such as textures or lighting conditions may be used to group together different objects; consequently, loading and unloading operations per object are now perform per group of objects.

Table 1 shows the features of five open standards (VRML, X3D, SMIL, SVG, MPEG-4) published by ISO or W3C for handling different aspects of 2D/3D graphics and scene graph.

Table 1. Comparative analysis of the multimedia open standards.

	VRML	**X3D**	**SMIL**	**SVG**	**MPEG-4**
Standard name	Virtual Reality Modelling Language	eXtensible 3D	Synchronised Multimedia Integration Language	Scalable Vector Graphics	Coding of audio-visual objects
Organisation	ISO	ISO	W3C	W3C	ISO
Publication Year	1997	2005	1998-2007	2003	1998-2007
Media types	3D graphics and ref. to video / audio	2D/3D graphics and ref. to video / audio	Ref. to 2D graphics, video / audio	2D graphics, text and image	Still image, video, audio, 2D/3D graphics
Animation	Linear Interpolation	Linear Interpolation	Local animation	Not relevant	Linear and higher order interpolation
Compression	No	Yes	No	No	Yes
Streaming	Only for video / audio	Only for video / audio	Only for video / audio	Not supported	Video, audio, scene graph, graphics animation
Synchronisation	Only between animation tracks	Only between animation tracks	Not supported per frame	Not relevant	Full support
Interactivity	Yes	Yes	Yes	Yes	Yes

The main target of VRML was the publishing of 3D models and worlds over Internet, by implementing similar functionalities as HTML does for text. Once the node names are known, VRML is very easy to author since it is based on a human readable description. As VRML limitations, let us mention the lack of compression and the non support of streaming for graphical contents. X3D improves VRML by using XML to formalise the data structure and implicitly reuses all the XML text processing tools as parsing, transformation, presentation, and compression. X3D is expected to be more widely adopted by the industry than VRML. However, X3D does not support streaming of graphical content which is an obstacle in many 3D graphics-based applications.

In the 2D graphics world, the W3C consortium published the recommendations for designing multimedia presentation within two standards: 1) SMIL which describes the temporal behaviour by synchronising the elementary media objects and 2) SVG which describes the 2D graphics primitives and their spatial properties. Both formalisms are XML-based, enabling easy integration and taking advantage of XML related tools. While the combination of both standards brings a rich solution in terms of features such as animation, content control, layout, linking, timing and graphics primitives, no support for streaming the graphic content is provided.

The MPEG-4 standard adopted a large part of the scene graph and graphics features presented in the standards mentioned above. The principle of MPEG-4 is to globally address the multimedia presentation issue and to specify requirements for a wide range of applications. However, the key advances of the MPEG-4 standard are the solutions it provides for the highlighted drawbacks of the above-mentioned standards: compression, streaming and 3D objects management.

The most differentiating contribution of MPEG-4 with respect to MPEG-1 and MPEG-2 is the definition of a format for multimedia scenes as part of the system information. This format lies on the top of all the media data and acts as an intermediate layer between media data and displayed content. This format provides a flexible way to manipulate various types of media in an MPEG-4 scene, allowing scheduling, control in temporal and / or spatial domain, synchronisation and management of interactivity.

Delivering MPEG-4 content may be achieved in three manners, depending on the targeted application:

1. Binary Format for Scenes (BIFS) is a collection of 2D and 3D nodes and their compression rules,
2. Binarized XML consists in applying a generic XML compression tool standardised by MPEG and called BiM on top of XMT (the textual description of MPEG-4 scene graph nodes),

3. Lightweight Application Scene Representation (LASeR), a subset of
   2D nodes, designed for addressing low performance terminals such as
   mobile phones.

Table 2 shows the main features of the three approaches.

Table 2. The different solutions for distributing MPEG-4 content and their respective characteristics.

	BIFS	XMT + BiM	LASeR
Targeted applications and terminals	Internet, Broadcasting, simple games on any terminal	Internet, Broadcasting, simple games on any terminal	Simple user interfaces: menus, EPGs on low performance terminals
Complexity	Moderate	Moderate	Low
Graphics Objects	2D and 3D	2D and 3D	2D
Basic graphics primitives	Box, Circle, Cone, Cylinder, Rectangle, Sphere, Text, Quadric	Box, Circle, Cone, Cylinder, Rectangle, Sphere, Text, Quadric	circle, rect, line, ellipse, text
Complex graphics representation tools	IndexedFaceSet, Curve2D, NURBS, Subdivision Surfaces, Solid Modelling	IndexedFaceSet, Curve2D, NURBS, Subdivision Surfaces, Solid Modelling	polygon, polyline
Support for Elementary Streams	Image, video, audio, animation and compressed 3D graphics	Image, video, audio, animation and compressed 3D graphics	Image, video, audio and animation

In contrast to LASeR which supports only 2D graphics, the BIFS approach
supports 2D and 3D graphics. In addition, BIFS can accommodate 3D
graphics compression streams, this feature being not yet available either in
XMT+BiM or in LASeR. For such reasons, MMW.com is based on BIFS.

BIFS is a binary encoded version of an extended set of VRML that is
more efficient than VRML in terms of data compactness. In addition to the
encoding schema, BIFS contains an enriched set of 2D and 3D graphics
primitives and most importantly, includes the mechanisms to stream the
graphical content.

To understand the role of BIFS within an MPEG-4 presentation let us review the architecture of an MPEG-4 terminal as presented in Fig. 2.

When receiving an MPEG-4 stream or when loading an MPEG-4 file, the first information read by the terminal is the *Initial Object Descriptor* (IOD). IOD includes references to one or two Elementary Stream (ES) Descriptors. If the MPEG-4 stream contains a scene, one of the ES descriptors refers to the *Scene Description Stream*. If it contains elementary media such as video, audio, still picture, compressed geometry or avatar animation, one of the ES descriptors refers to an *Object Descriptor Stream*. Note that several cases may occur: the file has a scene but no elementary media, the file has elementary media but no scene or the file has both a scene and elementary media stream.

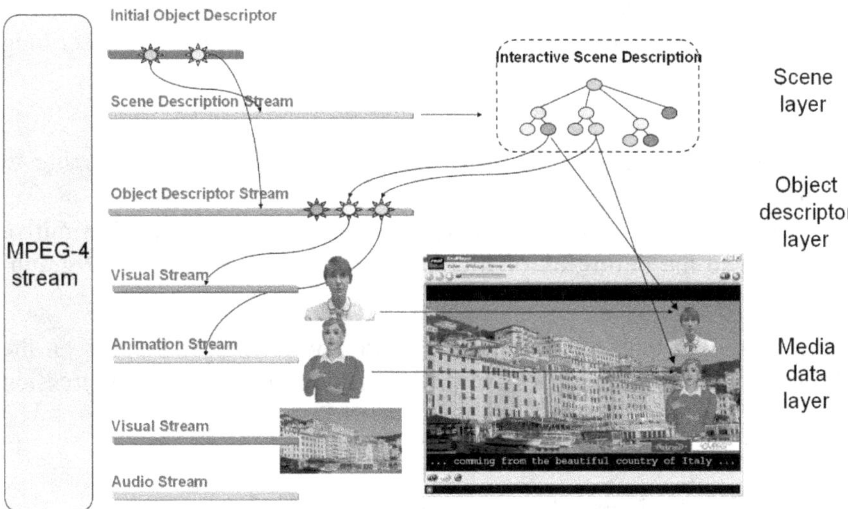

Fig. 2. MPEG-4 terminal data flow

When the IOD contains a reference to a *Scene Description Stream* (SDS), the SDS is decoded by the BIFS decoder and used to build the hierarchical structure represented as a tree in the terminal memory. Some branches of the tree contain references to media objects, others include information such as grouping, position in the rendered scene or state (activated or not, sensitive to user input or not). The tree structure is not necessarily static: it is possible to delete, replace and add new nodes. By enabling the server to perform these operations at any time, the MPEG-4 system supports streaming capabilities at the scene level: nodes, subgraphs or entire graphs can be received and the local scene can be dynamically updated. Within the MPEG-4 model, each media object has a local coordinate system. The media are positioned in the scene after the computation of their global position and orientation. Some parameters exposed by the nodes are accessible to the user by using the event propagation between sensors and media or graphics primitives. All the parameters referring to the object position or to different properties (form, colour) can be directly exploited by a search engine able to parse the MPEG-4 content.

The references to the media object are stored in a dedicated container called *Object Descriptor Stream*. Each descriptor refers to a media *Elementary Stream* and contains information related to it (resolution, memory buffer size, …). Here again, an indexing engine can formulate a query at this low information level.

A 3D graphics object is a complex structure, usually defined by a (scene) graph. It relies on nodes defining the geometry, the appearance and the animation. Similarly to generic nodes in the scene graph, 3D graphics nodes contain properties that may be parsed and interpreted for describing the content.

In MPEG-4, the support for 3D graphics is achieved at two levels:

- by defining specific nodes in the scene graph (each one referring to various types of graphics primitives) and,
- by applying compression schemes adapted to each kind of primitive (leading to the definition of Elementary Streams specific to 3D graphics).

The next section presents these 3D graphics primitives. It focuses on the MPEG-4 features allowing 3D online publishing, namely the compression and streaming functionalities.

2.3 The 3D graphics primitives in MPEG-4

The most common manner to represent 3D shapes in computer graphics is to approximate them by polygonal meshes. MPEG-4 uses the so-called IndexedFaceSet (IFS), as defined in VRM. IFS is defined as a set of vertices and a set of polygons connecting the vertices. Additionally, MPEG-4 introduces a dedicated compression schema, called 3DMC (3D Mesh Compression), allowing to reduce the size of the file with ratios of up to 40:1 [23]. 3DMC supports advanced streaming capabilities such as incremental transmission and rendering, progressive coding by Progressive Forest Split [31] and error resilience [23]. Here, the object geometry is represented in an unstructured manner, with very few semantics. Any shape may be represented as an IFS. However, information such as the number of vertices, polygons, the existence of texture may be easily extracted from the IFS and then used for describing the content. Additionally 3D shape information can be provided by applying to the 3D mesh representation the 3D/2D shape descriptors developed in the literature [4, 33, 34, 35]. The efficient API we developed demonstrates how easily MMW.com can support them.

Beside IFS and its compressed version, MPEG-4 supports also higher order approximations as Non-Uniform Rational B-Splines (NURBS) [22, 25] and Subdivision Surfaces (SS) [22]. Subdivisions are built on the top of coarse, low resolution meshes expressed with IFS and follow the rules of well-known schemes as Loop's [22], and Catmull–Clark's [5]. When dealing with the NURBS representation, the surface order may become a relevant descriptor. When considering SS, in addition to descriptors associated with IFS, the subdivision schema may also be a useful descriptor directly available.

Wavelet subdivision surfaces support the addition of 3D details at each step of the subdivision process. Therefore, the descriptors directly obtained from this representation are related to the compressed bitstream and may be the number of vertices per additional layer and mapping of the geometric regions in the bitstream.

MPEG-4 also supports the solid modelling [27] which enables the content authors to create complex volumes using an exact geometry. While the previously introduced geometric models describe the surface of an object, solid models describe the volume, thus enabling operations on densities. In such a case, the coefficients of implicit equations carry out very few semantics, and more advanced algorithms are needed to describe the shape of such 3D objects.

Most of the MPEG-4 tools for 3D graphics compression natively support streaming. Data are partitioned and continuously transmitted so that the user does not have to wait for the entire content before rendering. This feature, together with the support of audio/video contents makes MPEG-4 a good candidate for online publishing of multimedia data. Therefore, the internal format of MMW.com for data storage as well as for content publishing relies on MPEG-4.

In the next section we show how to enrich the MPEG-4 content with MPEG-7 and proprietary descriptors, either by directly parsing the MPEG-4 scene graph or by providing specific media analysis tools.

3. www.MyMultimediaWorld.com: an online, open standard-based MAMS

The relevance of the MPEG-7 descriptors was already shown in audio-visual applications, by adding MPEG-7 descriptions to videos[6] (annotations of scenes, temporal segmentation, key images...) [8, 18] or to images [5, 6]. However, the image/video servers currently available in free access over the Internet such as Flickr[7] or YouTube[8] only support text-based indexation of the content, exploiting user annotations and/or associated tags.

One of the main difficulties in indexing multimedia content is the availability of algorithms and software components to extract meaningful descriptors. While defining a complex and almost complete schema for audio/video content description, MPEG-7 does not specify methods for descriptor extraction. Moreover, descriptor diversity and complexity make the implementation of a multi-descriptor MAMS by a single contributor almost impossible. In order to support software modules for indexing and retrieval purposes, we developed MMW.com which is an online MAMS based on MPEG-4 and MPEG-7 (for the content representation and description, respectively). In MMW.com the complex problem of providing metadata is solved by implementing an open API, which allows to plug-in *third party* components:

- to propose schema extensions by enriching the MPEG-7 XSD,
- to plug-in description extractors that fill the sub-parts of the schema defined in MPEG-7 or not,
- to plug-in distance measurement software applied to the schema sub-parts.

The API should be agnostic to the data type. Since the MPEG-7 features for audio and video content are extensively described in other chapters of this book, we only focus on the MPEG-7 features for 3D graphics, and we test this property by implementing all the above-mentioned three functionalities for 3D graphic content.

The MPEG-7 [20] defines two 3D shape descriptors, namely Shape Spectrum and Perceptual Descriptor. However, there are very few examples of applications using MPEG-7 for 3D content description. On the one hand, research activities are focused on the MPEG-7 descriptor performances for indexing and retrieval purposes [15, 33, 34, 35], and on the other hand on 3D scene representation [3]. However none of them proposed yet

---

[6] MovieTool. www.ricoh.co.jp/src/multimedia/MovieTool/index html
[7] www.flickr.com
[8] www.youtube.com

an MPEG-7 (or extended MPEG-7) description for an online 3D graphics assets management system. Such a system should provide a fast and consistent retrieval functionality of multimedia objects in large databases.

Originally, the MPEG-7 Schema was not designed to offer web service functionalities, since it refers mainly to elements related to the multimedia content itself. We appended the MPEG-7 XSD with the elements specific to online distribution such as date of upload in the library, visualisation number, and user community. Such information is not directly extracted from the content itself but could be attached to it, enriching the semantic knowledge on the considered media. Therefore we appended the MPEG-7 XSD instead of creating a specific XSD for content distribution and user community.

The second extension we implemented aims at supporting a better indexation of 3D graphics objects by using the basic elements directly extracted from the MPEG-4 graphics primitives: number of vertices, existence of a texture map and coding type. The complete description of this extension, called Properties3DObjectType, is illustrated in Fig. 3.

```
<complexType name="Description3DObjectType">
 <complexContent>
  <extension base="mpeg7:StillRegion3DType">
   <sequence>
    <element name="Properties3DObject"
             type="mod:Properties3DObjectType" minOccurs="0"/>
   </sequence>
  </extension>
 </complexContent>
</complexType>
<complexType name="Properties3DObjectType">
  <sequence>
    <element name="NumberOfVertex" type="int" minOccurs="0"/>
    <element name="NumberOfComponents" type="int" minOccurs="0"/>
    <element name="NumberOfTriangles" type="int" minOccurs="0"/>
    <element name="CodingType" type="string" minOccurs="0"/>
    <element name="HasTexture" type="boolean" minOccurs="0"/>
    <element name="isManifold" type="boolean" minOccurs="0"/>
    <element name="isAnimated" type="boolean" minOccurs="0"/>
    <element name="BitsPerVertexBIFS" type="float" minOccurs="0"/>
    <element name="BitsPerVertex3DMC" type="float" minOccurs="0"/>
    <element name="TextureSize" type="float" minOccurs="0"/>
    <element name="OriginalFormat"
             type="ControlledTermUseType" minOccurs="0"/>
    <element name="BasicGraphicsPrimitives"
             type="mmw:BasicGraphicsPrimitivesType"
             minOccurs="0" maxOccurs="1"/>
    <element name="NURBSOrder"
             type="int" minOccurs="0" maxOccurs="1"/>
    <element name="SubdivisionSchema"
             type="int" minOccurs="0" maxOccurs="1"/>
  </sequence>
</complexType>
```

```
<complexType name="BasicGraphicsPrimitivesType">
  <sequence>
    <element name="Box" type="mmw:YesNoSelectionType"
             minOccurs="0" maxOccurs="1"/>
    <element name="Circle" type="mmw:YesNoSelectionType"
             minOccurs="0" maxOccurs="1"/>
    <element name="Cone" type="mmw:YesNoSelectionType"
             minOccurs="0" maxOccurs="1"/>
    <element name="Cylinder" type="mmw:YesNoSelectionType"
             minOccurs="0" maxOccurs="1"/>
    <element name="Rectangle" type="mmw:YesNoSelectionType"
             minOccurs="0" maxOccurs="1"/>
    <element name="Sphere" type="mmw:YesNoSelectionType"
             minOccurs="0" maxOccurs="1"/>
    <element name="Text" type="mmw:YesNoSelectionType"
             minOccurs="0" maxOccurs="1"/>
    <element name="Quadric" type="mmw:YesNoSelectionType"
             minOccurs="0" maxOccurs="1"/>
  </sequence>
</complexType>
```

Fig. 3. Example of a proprietary type, extending an MPEG-7 one and containing 3D object basic descriptors extracted from the MPEG-4 data.

In the next section, we describe the API functional implementation. It was designed so as to let the system open for new user defined descriptors. Moreover, it meets the third party requirement as defined above, *i.e.* the possibility to extend the MPEG-7 schema, to compute the descriptors themselves as well as the distances between them.

3.1 MMW.com API description

The developed API aims at a user-friendly and transparent media access with respect to the MPEG-4 compression algorithms. In this respect, very simple data structures, easy to be parsed, are defined for each media type: bitmap for images and videos, vertices and index buffers for 3D graphics. For the third party contributors, the main advantages of our system are connected to its flexibility and the existence of a database: based on this API, he/she may implement his/her own algorithms in a library organised as illustrated in Fig. 4. Also note that this API makes it possible to benchmark different algorithms achieving descriptor extraction in any media file.

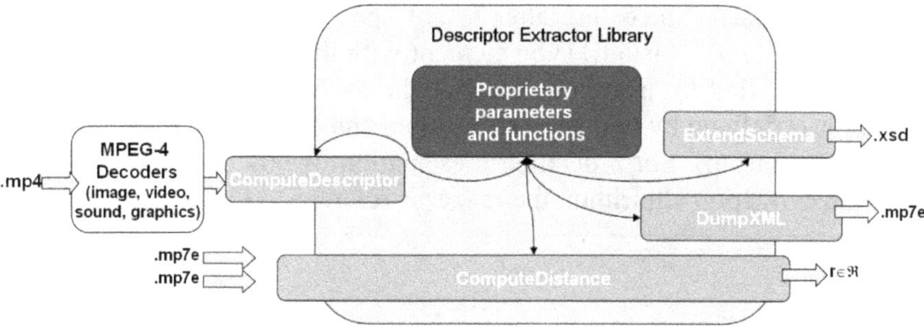

Fig. 4. Structure of the third party contributor library for descriptor extraction using the API (.mp7e denotes an extended MPEG-7 file)

The MMW.com API has four abstract functions ensuring the following functionalities. The most important are:

- **Extension of the MPEG-7 XSD:** This component, called Extend-Schema, specifies the user defined XSD. The system includes the proprietary description schema in the MMW.com schema by using the *include* mechanism of XML[9]. The MPEG-7 type extended by this component should also be known: the proprietary-defined XSD should import the MPEG-7 XSD[10].
- **Description extraction:** This component, called ComputeDescriptor, has as input the data structures obtained by decoding the MPEG-4 content from the database. The output is an instantiation of the MPEG-7 (extended) schema in function DumpXML.
- **Distance calculation:** Starting from two MPEG-7 (extended) instantiations, this component, called GetDistance, provides a distance represented as a real number.

To illustrate the API and its appropriation by a user a programming template (Visual C++ project) is providing online[11].

3.2 API implementation for 3D graphics

Based on the previously described API, we implemented two libraries of 3D shape descriptors. The first one, called Shape Spectrum is part of the MPEG-7 standard. The second one, based on the Hough Transform [33,

---

[9] vega.int-evry.fr/3dod_php_dev/MMWSchema/MMWSchema.xsd
[10] vega.int-evry.fr/3dod_php_dev/MMWSchema/MMWSchema_Hough3D_v1.xsd
[11] vega.int-evry.fr/3dod_php_dev/MMW_API/MMWLib.zip

34] is proprietary. Since the latter is not specified in MPEG-7, we extended the MPEG-7 VisualDType element with the type presented in Fig.5 which is specified by using the ExtendSchema function of the API. This descriptor is defined by two parameters (step and levels) and by an array of doubles (Spectrum). For a detailed presentation of this descriptor, as well as for the extraction algorithm, the reader is referred to [33].

```
<complexType name="Hough3DType">
  <complexContent>
    <extension base="mpeg7:VisualDType">
      <sequence>
        <element name="Spectrum" type="mpeg7:doubleVector"/>
        </sequence>
      <attribute name="step" type="mpeg7:unsigned8"
                 use="optional" default="20"/>
      <attribute name="levels" type="mpeg7:unsigned8"
                 use="optional" default="2"/>
    </extension>
  </complexContent>
</complexType>
```

Fig. 5. Proprietary defined XSD for the 3D Hough descriptor

In order to extract the Shape Spectrum and 3D Hough descriptors we implemented the ComputeDescriptor and GetDistance functions of the API.

3.2 Result visualisation on www.MyMultimediaWorld.com

MMW.com supports several mechanisms for searching the content. The first one parses the textual annotations attached to an MPEG-4 file: title, tags, description introduced by the owner, comments provided by other users… The second one consists in query by example. When visualising a file, the system presents to the user the "semantic search" option if the media has associated MPEG-7 (extended) data. If "semantic search" is selected, the system shows similar content. Similarity is obtained as a list of closest objects with respect to the distance computed by the descriptor extractor libraries. For each descriptor, the closest objects obtained are showed in a row. Fig. 6 (top) illustrates the web-based presentation engine by an example of retrieval based on the 3D Hough and Shape Spectrum descriptors.

Additionally, the retrieval results may be filtered with respect to properties that are obtained by directly parsing the MPEG-4 data. As an example (Fig. 6, bottom), it is possible to search for 3D objects with more or less the same number of vertices (a percentage controlled by the user), with or without texture, manifold or not, animated or not.

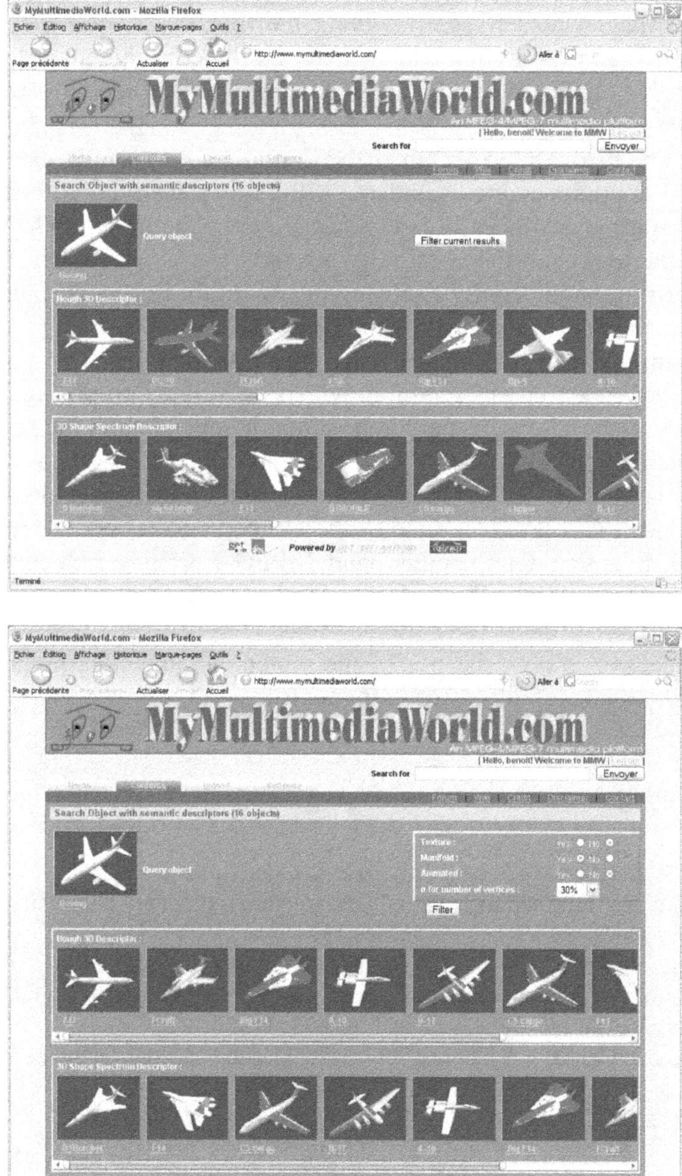

Fig. 6. Results of a query by example: 3D objects are indexed with the 3D Hough and the Shape Spectrum descriptors (top) and then additionally filtered (bottom)

4 Conclusion

In this chapter we introduced MMW.com an advanced MAMS (Multimedia Asset Management System) solving the four key issues in the field: unified converters for input media, generic multimedia content descriptors, user-friendly interaction, and integrated image/audio/video/3D presentation engine.

At our knowledge MMW.com is the first attempt to jointly use two international multimedia open standards MPEG-4 (content compression and representation) and MPEG-7 (content description) for an online MAMS. It provides the user with the possibility not only to upload content but also update software modules. By exposing interfaces for user contributions for enriching the performances/capabilities of the description schema, MMW.com goes one step further than current Web2.0 applications. Now, users may take the role of solution providers as well, by proposing new descriptors or description scheme that is smoothly integrated in the system.

We thus demonstrate that the combined use of MPEG-4 and MPEG-7 standards is an efficient solution for online, multimedia MAMS. On the one hand, MPEG-4 provides state of the art media compression, representation and streaming. On the other hand, MPEG-7 brings a rich content description schema and mechanism to extend it further by user defined extensions.

Our objective is now the use of the system as an online platform for benchmarking image/video processing tools.

References

1. Adali S, Candan KS, Chen SS, Erol K, Subrahmanian VS (1996) The advanced video information system: Data structures and query processing. ACM-Springer Multimedia Systems Journal:172 – 186
2. Behrendt W, Fiddian NJ (2001) Database interoperation support in multimedia applications: architecture and methodology. Electronics & Communication Engineering Journal, vol 13, issue 4, pp 173 – 182
3. Bilasco IM, Gensel J, Villanova-Oliver M, Martin H (2005) On indexing of 3D scenes using MPEG-7, ACM MM'05, Singapore, pp 471 – 474
4. M. Bober, F. Prêteux, W.-Y. Kim (2002), Shape Descriptors, *in* B.S. Manjunath, P. Salembier, T. Sikora (Ed.), Introduction to MPEG-7: Multimedia Content Description Language, John Wiley & Sons, New York, NY, pp 231 – 260
5. Catmull E, J Clark (1978) Recursively generated B-spline surfaces on arbitrary topological meshes, Computer-Aided Design, vol 10, pp 350 – 355

6. Chang SF, Chen W, Sundaram H (1998) VideoQ: a fully automated video retrieval system using motion sketches, Fourth IEEE Workshop on Applications of Computer Vision, pp 270 – 271

7. Chang W, Yoon K, Gruhne M, Villegas P (2007) MPEG-7 Query Format Requirements, *ISO/IEC JTC1/SC29/WG11*, MPEG07/N8780, Marrakech

8. Ching-Yung L, Tseng BL, Smith JR (2002) Universal MPEG Content Access Using Compressed-Domain System Stream Editing Techniques, IEEE Int'l Conf. on Multimedia and Expos, Lausanne, vol 2, pp 73 – 76

9. Funkhouser T, Kazhdan M (2005) Shape-based Retrieval and Analysis of 3D Models, ACM SIGGRAPH, Los Angeles, vol 48(6), pp 58 – 64

10. Funkhouser T, Min P, Kazhdan M, Chen J, Halderman A, Dobkin D, Jacobs D (2003) A Search Engine for 3D Models, ACM Transactions on Graphics, vol 22(1), pp 83 – 105

11. Hacid MS, Decleir C, Kouloumdjian J (2000) A database approach for modeling and querying video data, IEEE Transactions on Knowledge and Data Engineering, vol 12(5), pp 729 – 750

12. Information technology -- Coding of audio-visual objects -- Part 11: Scene description and application engine, ISO/IEC 14496-11:2005

13. Johnson RB (1999) Internet multimedia databases, IEE Colloquium on Multimedia Databases and MPEG-7, pp 1 – 7

14. Kalipsiz O (2000) Multimedia databases, IEEE International Conference on Information Visualization, London, pp 111 – 115

15. Kim DH, Park IK, Yun ID, Lee SU (2004) A New MPEG-7 Standard: Perceptual 3-D Shape Descriptor, Lecture Notes in Computer Science, vol 3332/2004, pp 238 – 245

16. Kiranyaz S, Caglar K, Guldogan E, Guldogan O, Gabbouj M (2003) MUVIS: a content-based multimedia indexing and retrieval framework, Seventh International Symposium on Signal Processing and Its Applications, vol 1, pp 1 – 8

17. Kiranyaz S, Gabbouj M (2007) Hierarchical Cellular Tree: An Efficient Indexing Scheme for Content-Based Retrieval on Multimedia Databases, IEEE Transactions on Multimedia, vol 9(1), pp 102 – 119

18. Klamma R, Spaniol M, Jarke M (2005) MECCA: Hypermedia Capturing of Collaborative Scientific Discourses about Movies, informing science. The International Journal of an Emerging Discipline, N. Sharda (ed): Special Series on Issues in Informing Clients using Multimedia Communications, vol 8, pp 3 – 38

19. Lux M, Becker J, Krottmaier H (2003) Caliph & Emir: Semantic Annotation and Retrieval in Personal Digital Photo Libraries, Proceedings of CAiSE '03 Forum at 15th Conference on Advanced Information Systems Engineering, Velden, pp 85 – 89

20. Manjunath BS, Salembier P, Sikora T (2002) Introduction to MPEG 7: Multimedia Content Description Language, Wiley (ed)

21. Mezaris V, Doulaverakis H, Herrmann S, Lehane B, O'Connor N, Kompatsiaris I, Strintzis MG (2005) The SCHEMA Reference System: An Extensible Modular System for Content-Based Information Retrieval, Workshop on Image Analysis for Multimedia Interactive Services (WIAMIS), Montreux

22. MPEG 3DGC Homepage [Online]. Available: http://www mpeg-3dgc.org/
23. MPEG: "ISO/IEC 14 496/Amd 1 Extensions" (a.k.a. "MPEG-4 Version 2"), ISO/IEC, 2000.
24. Pentland A., Picard R, Sclaroff S (1994) Photobook: Content-based Manipulation of Image Databases, In SPIE Storage and Retrieval for Image and Video Databases II, San Jose
25. Pieg L, Tiller W (1997) The NURBS Book., Springer-Verlag (eds), Berlin
26. Ribeiro C, David G, Calistru C (2004) A multimedia database workbench for content and context retrieval, IEEE 6th Workshop on Multimedia Signal Processing, pp 430 – 433
27. Rotgé J-F (1996) Principles of solid geometry design logic, in Proc. CSG Conf, Winchester, pp 233 – 254
28. Salomie A., Deklerck R, Munteanu A, Cornelis J (2002) The MeshGrid Surface Representation. Vrije Universiteit Brussel, Dept. ETRO-IRIS, Tech. Rep. IRIS-TR-0082
29. Santini S, Gupta A (2002) Principles of schema design for multimedia databases, IEEE Transactions on Multimedia, vol 4(2), pp 248 – 259
30. Smith JR, Chang SF (1996) Visualseek: a fully automated content-based image query system, In Proceedings of ACM Multimedia 96, Boston, pp 87-98
31. Taubin G, A Guéziec, Horn W, Lazarus F (1998) Progressive forest split compression, in Proc. ACM SIGGRAPH, pp 123 – 132
32. Yoon K, Doeller M, Gruhne M, Tous R, Sano M, Choi M, Lim TB, Lee J.J, Seo HC (2007) MPEG-7 Query Format, *ISO/IEC JTC1/SC29/WG11*, MPEG07/N9103, USA
33. Zaharia T, Preteux F (2001) Hough transform-based 3D mesh retrieval, Proceedings of SPIE, the International Society for Optical Engineering, vol 4476, pp 175 – 185
34. Zaharia T, Prêteux F (2004) 3D versus 2D/3D shape descriptors: A comparative study Proceedings of SPIE Conference on Image Processing: Algorithms and Systems III - IS&T / SPIE Symposium on Electronic Imaging, Science and Technology '04, San Jose, CA, Vol. 5298, pp 47 – 58
35. Zaharia T, Prêteux F (2001) 3D-shape-based retrieval within the MPEG-7 framework Proceedings of SPIE Conference on Nonlinear Image Processing and Pattern Analysis XII, San Jose, CA, Vol. 4304, pp 133 – 145

Overview of Open Standards for Interactive TV (iTV)

Günther Hölbling, Tilmann Rabl and Harald Kosch

Department of Distributed Information Systems, University of Passau
Innstrasse 43, 94032 Passau, Germany
{guenther.hoelbling, tilmann.rabl, harald.kosch}@uni-passau.de

Television has become the most important mass medium. The digitalisation in progress gives new possibilities to enrich the television experience. The Electronic Program Guide - though still very limited - gives a first impression of future television. There are various attempts to give the audience a more active role. All these can be found under the topic interactive TV. Besides the world of television, there is a fast growing community for videos in the World Wide Web (e.g. YouTube[1] or MyVideo[2]). Until today interactivity in "WWW-videos" is of minor importance, but has a great potential for growth. It has not reached a similar level of professionalism as in iTV and no open standards have been introduced. Thus we will concentrate on interactivity in traditional TV. In this chapter, we present different forms of interaction in television and give an extensive overview of existing platforms and standards like MHEG, DAVIC, Java TV, MHP, GEM, OCAP and ACAP. Also the underlying technology that forms the basis for these standards is presented in short. Finally, the relationships between the different standards and a look at future trends of iTV will be given.

1 Introduction

The term "interactive television" (iTV or ITV) is used for television systems in which the audience may interact with the television content. Interactivity in TV is usually not intended to mean that the viewer is enabled to change the storyline of a program. Rather he/she may participate in a quiz show, gather additional information on news topics or directly buy a product presented in a commercial. Also Electronic Program Guides (EPGs), Video on Demand (VoD) Portals or Telelearning can be realized with iTV Systems.

---

[1] YouTube - http://www.youtube.com/
[2] MyVideo - http://www.myvideo.de/

G. Hölbling et al.: *Overview of Open Standards for Interactive TV (iTV)*, Studies in Computational Intelligence (SCI) **101**, 45–64 (2008)
www.springerlink.com

Interactivity in television is not as new as one might think. In the beginning, around 1960, there were shows, where viewers called in and participated in quiz shows. In 1974, the teletext was developed in the United Kingdom. It was the first form of additional content delivered with the broadcast programs. At the IFA[3] 1979 the Tele-Dialog was introduced. It was a televoting system which allowed viewers to participate in polls for TV shows by calling specific telephone numbers. It was not until the 1990s that more advanced forms of interactivity appeared in the media landscape. One of the first trials was the Full Service Network by Time Warner, which was launched in December 1994. This trail provided several interactive services, like video-on-demand, program guide, video games, and home-shopping to the customers. In 18 months only 65 people subscribed and therefore it was closed again. With the ongoing digitalisation of television, interactivity became an interesting topic again, and many broadcasters have been trying to enrich their product range by enhancing it with interactive services.

We start with a categorisation of interactivity in TV (section 2). iTV systems are complex systems that involve a long chain of successive processes from the broadcaster to viewers. Section 3 gives a short overview of the main components of such systems. One of the most interesting and important components to empower the viewer to use iTV applications, the middleware and different iTV standards are discussed in section 4. The last section concentrates on the interrelation between different standards.

2 Levels of Interactivity

Interactive applications have a widespread diversity of user interfaces and resource requirements but also different levels of interactivity. In this section we will introduce seven levels and for each level we will give some representative interactive applications. Our categorisation is based on [26].

Level 1 - Basic TV: Interaction on this level is defined as basic functionalities for watching TV such as switching channels or powering the TV set on and off.

Level 2 - Call-In-TV: At this level, the interaction between the audience and the broadcaster is established by techniques such as telephone calls or short message service (SMS). Examples of such TV shows are televoting shows where viewers can vote for a candidate or the next music clip.

Level 3 - Parallel TV: Parallel TV introduces alternative content on multiple channels. The viewer is empowered to change the way he/she consumes a broadcast program. Popular examples for broadcasts on this level are multilingual audio channels or subtitles. Another form of parallel TV are shows with different camera angles, well known from auto racing pro-

---

[3] Internationale Funkausstellung Berlin

grams. A very special form are movies that show the perspective of different characters on different channels.

Level 4 - Additive TV: This level is also known as "enhanced TV". Additionally to the TV program further content is broadcast. The content can be basic information or advanced services. A well-established service is teletext. Applications like EPG or synchronised computer programs are advanced examples. A return channel is not needed for applications on this level.

Level 5 - Service on Demand: The "Media on Demand" level enables the viewer to consume programs detached from the TV schedule. This level includes video on demand (VoD), upgrade services and other services that are provided on demand. The interaction between the user and the service provider requires a return channel. In TV environments, VoD is not very common because of its technical demands. Often, Near-VoD is used instead of VoD. Near-VoD uses several channels where multiple copies of a program are broadcast in short intervals.

Level 6 - Communicative TV: For such programs, content from other sources such as the Internet can be accessed in addition to broadcast content. Services that stem from the PC domain can also be used and combined with TV. At this level, TV may also be enriched by community functions: chats, online games or email. Another option is user-generated content that can be uploaded. Thus, user- or community-generated programs become possible.

Level 7 - Fully Interactive TV: The most enhanced level of interactivity enables the user to create her/his individual storyline for a program. A program on this level can be understood as a kind of video game, in which the user affects the proceeding of the program. The program can also be affected automatically based on personal profiles, which may include personalised commercials as well as personalised movies.

Today most iTV applications can be assigned to the level 4, 5 or 6. Nevertheless user-generated content, as mentioned on level 6, is hardly present in the world of iTV. Applications on level 7 are still in their infancy although several approaches for personalised commercials are in progress (e.g. Advertising that is relevant to a person[4]).

3 Basic Technologies for iTV

Before discussing the different iTV standards, we will give a brief introduction of the basic technology for iTV.

---

[4] United States Patent Application: 20070174117; Advertising that is relevant to a person by the Microsoft Corporation

3.1 Set-top Boxes

A set-top box (STB) is a device that forms a link between a TV-set and an external source[5]. The source usually is one of the following: cable, terrestrial antenna or satellite dish. The term set-top box stems from the fact that STBs usually are placed above the TV-set. STBs have multiple purposes, ranging from simple signal conversion (e.g. digital receiver) over personal video recorder functions to interactive media center functions.

3.2 Video Encoding and Transport

Audio and video encoding standards play a major role in the world of digital TV (DTV). The encoding and the resulting reduction of data leveraged the introduction of Digital TV. An important standard which is pervasive in DTV is the MPEG-2 standard by the Moving Picture Experts Group[6]. Also MPEG-4 Part 10 Advanced Video Codec (MPEG-4 AVC or H.264) is used in several recent DTV standards. Besides the video and audio encoding capabilities of MPEG-2, the system part of the standard is of significant importance in DTV. The MPEG-2 system part describes the combination of multiple encoded audio and video streams and auxiliary data into a single bitstream. This makes it possible to carry multiple TV channels, additional data and iTV applications.

3.3 Digital TV Standards

The term "Digital TV" (DTV) or "Digital Video Broadcasting" (DVB) is used for the transmission of digitised audio, video and auxiliary data. As for analog broadcasting, the digital TV market is fragmented. There are various standards developed by different organisations all over the world. This fragmentation is even extended by the accommodation of the DTV standards to different transmission channels used for broadcasting. For that reason, there are different standards for terrestrial, satellite, cable and mobile TV. In Europe, the DTV standards of the Digital Video Broadcasting Project[7] (DVB) are used. Other major players in the development of DTV standards are the Advanced Television Systems Committee[8] (ATSC), the CableLabs[9] in the US and the Association of Radio Industries and Businesses[10] (ARIB) in Japan.

---

[5] According to more general definitions all forms of electronic devices that are connected to a TV-set are called set-top boxes, but we will use the more restricted definition from now on.

[6] The Moving Picture Experts Group (MPEG) - `http://www.chiariglione.org/mpeg/index.htm`

[7] Digital Video Broadcasting Project - `http://www.dvb.org/`

[8] Advanced Television Systems Committee (ATSC) - `http://www.atsc.org/`

[9] CableLabs - `http://www.cablelabs.com/`

[10] The Association of Radio Industries and Businesses (ARIB) - `http://www.arib.or.jp/english/`

4 Middleware Platforms

In this section we will give an overview of several open iTV standards and middleware platform specifications. This overview is not exhaustive, but covers the major open standards.

4.1 MHEG

The Multimedia and Hypermedia Information Coding Expert Group[11] (MHEG), a subgroup of the International Organisation for Standardisation[12] (ISO), published the MHEG standard in 1997. It was designed as an equivalent to HTML for multimedia presentations [22]. Therefore, the aim of the group was to describe the interrelation between different parts of a multimedia presentation and to provide a common interchange format. The standard initially consisted of five parts, and three parts were added later on:

MHEG-1: MHEG Object Representation Base Notation (ASN.1) [15].
MHEG-2: Should provide an encoding based on SGML instead of ASN.1 but was never finished [25].
MHEG-3: MHEG Script Interchange Representation [16].
MHEG-4: MHEG Registration Procedure [14].
MHEG-5: Support for Base Level Interactive Applications [17].
MHEG-6: Support for Enhanced Interactive Applications [19].
MHEG-7: Interoperability and Conformance Testing for MHEG-5 [20].
MHEG-8: XML Notation for MHEG-5 [21].

In contrast to other standards, MHEG was designed to be only a description language for final-form interactive multimedia presentations. It provides neither a definition of a graphical user interface nor any architectural specification of the execution engine.

The first part of the standard defines the encoding of MHEG presentations in ASN.1 notation. A central aspect of the design was to build a generic standard. As such, it contains no specification about the application area or target platform. MHEG follows an object-oriented approach. Media elements of a presentation, such as text, audio and video, are represented by *Content* objects. These can contain information about the media elements, such as spatial and temporal attributes, as well as information about the actual content or a reference. *Action, Link,* and *Script* objects are used to describe behaviours. Simple objects can be grouped to *Composite* objects. The reason for this is to bundle objects that are needed together. Limited user interactivity is provided by the *Selection* and *Modification* class. The *Selection* class enables the user to select an item of a predefined set as in drop-down menus,

---

[11] Multimedia and Hypermedia Information Coding Expert Group (MHEG) - `http://www.mheg.org`
[12] International Organisation for Standardisation (ISO) - `http://www.iso.org`

while the *Modification* class provides free user input. For further information about these and the remaining classes please refer to the MHEG standard [15].

Part 2 was planned to provide an encoding based on SGML, but it was never finished.

The third part of the MHEG standard defines a virtual machine and an encoding model for scripting. The interactive elements of MHEG-1 are very limited. This part of the standard features advanced operations. The encoding is based on the Interface Definition Language of CORBA and is known as the Script Interchange Representation [9]. As for the other standard parts, there were no specifications made about the concrete execution environment. The encoding is mainly a intermediate representation for the platform-independent exchange of scriptware.

MHEG-4 describes the procedure to register identifiers for objects, for example for data formats or script types.

MHEG-1 and MHEG-3 had many features that were too complicated for the technology of their time. In order to overcome this problem and to support systems with minimal resources, MHEG-5 was developed. Although MHEG-5 is a simplification of MHEG-1, there are too many differences for the standards to be compatible. The class hierarchy itself is quite different. To reflect the interactive television domain, the naming was adapted. An MHEG-5 application consists of *Scenes*, that are composed of *Ingredients*.

MHEG-5, like MHEG-1, lacks possibilities for advanced processing of hypermedia data. Although MHEG-1 was already extended by the definition of a new script encoding standard and an according virtual machine in MHEG-3, the new standard MHEG-6 was specified. In contrast to MHEG-3, MHEG-6 builds upon existing solutions for data processing. MHEG-6 uses the Java programming language as a basis and defines an interface for the interoperability of MHEG and Java objects, the MHEG-5 API.

MHEG-7 defines a test suite for interoperability and conformance testing for MHEG-5 engines. Additionally, a format for test cases is defined to allow more detailed or application specific tests.

MHEG-8 defines an alternative encoding format for MHEG-5 objects based on XML.

Although MHEG-1 was not able to gain wide acceptance in the iTV market the reduced specification of MHEG-5 is used in several systems. Today MHEG-5 has great industry support and MHEG-5 content (MHEG 5 UK Version 1.06 [3]) is broadcast in the UK and New Zealand. Also many extensions and profiles such as Euro MHEG have been developed and are in use today.

4.2 DAVIC

The Digital Audio-Video Council[13] (DAVIC) was founded in 1994 and completed its work in 1999. It was a non-profit organisation with a membership of

---

[13] The Digital Audio-Video Council (DAVIC) - `http://www.davic.org/`

over 220 companies. Its purpose was to promote interactive digital audio-visual applications and to maximise interoperability across countries and applications and services [6]. Since there were so many companies involved, a major target was to keep the specifications to a minimum. So, existing standards were used whenever possible, and new ones created only if none existed. For example for multimedia information delivery the MHEG-5 format was used. The versions 1.0 - 1.4 of the standard are a set of 11 (v1.0) to 14 (v1.4.1) parts that cover all areas of commercial interactive multimedia experience. Version 1.5 is an additional set of five parts that especially pay attention to IP-based audio-visual services. After 5 years, the DAVIC work was completed. Some concepts and parts of DAVIC were resumed by the TV-Anytime[14] organisation. Nowadays, not all of the DAVIC specifications are used, but major parts are referenced in many other standards.

4.3 Java TV

The Java TV API[15] is an extension of the Java platform. It provides functionality for using Java to control and run applications on TV receivers, such as set-top boxes. The main purpose of these extensions is to combine the platform independence of Java applications with a set of functions recommended for an iTV platform offered by TV-specific libraries. Furthermore, Java TV applications are independent of the underlying broadcast network technology. The JVM resides on the set-top box and allows a local execution of the applications, usually embedded in the broadcast content. Set-top boxes are often very resource-constrained devices. For that reason, the PersonalJava application environment, which is optimized for such devices, is used. PersonalJava offers a subset of the APIs introduced by the Java Standard Edition. PersonalJava applications are fully compliant with the Java standard edition. There are several packages of the PersonalJava application environment that are often used by Java TV applications. The *java.io* package is for input/output operations. It is used for file-based operations such as filesystem access (local and remote) and for stream-based operations such as broadcast data access. The *java.net* package is used for IP-based network access. These functions are often used for providing return channels or accessing IP data in the MPEG TS. Another important feature is security. For this purpose, Java TV makes use of the JDK 1.2 security model which lets operators define their own security model or policy. Most important for an iTV system, in terms of security, are areas like the conditional access sub-system, secured communication and the secure execution of code in the JVM. Based on the graphics toolkit, the abstract window toolkit (AWT) offered in the *java.awt* package, user interfaces (UI) can be build. AWT offers a set of basic UI components.

---

[14] The TV-Anytime Forum (`http://www.tv-anytime.org/`) developed open specifications for metadata with a focus on all participants in the TV value chain.
[15] The Java TV API - `http://java.sun.com/products/javatv/`

In the following several important aspects and functions of Java TV will be explained [5].

Service and Service Information (javax.tv.service)

A service is often used as a synonym for "TV channel". In Java TV, a service is handled as a unit. It represents a bundle of content (audio, video and data) that can be selected by the user and presented on the receiver. Service information (SI) describes the content and layout of a service. This information can be offered in several formats depending on the used standard such as DVB-SI or ATSC A56. Java TV offers a common API for accessing the service information. Services are composed by several service components. A service component represents one element of a service such as a video or a Java application. Besides these functions that are common to all services, more specialised features are available. The *navigation* subpackage offers classes to navigate through the existing services and to request detailed information about services and their components. The *guide* subpackage provides APIs for EPG. Basic EPG features like program schedules, program events, and rating information such as information for parental control are also included. The *transport* subpackage offers additional information about the transport mechanism that is used, such as MPEG-2 TS. The *selection* subpackage provides mechanisms to select discovered services for presentation. The service context, represented by the *ServiceContext* Class, provides an easy way for controlling the service and its presentation. The selection of a service context changes the presentation of the service and its components. For example a selection may cause the receiver to tune in to a desired service, demultiplex the necessary service components, present the audio and video, and launch any associated applications. A service context may exist in one of the following four states - "Not Presenting", "Presentation Pending", "Presenting", "Destroyed". Although the number of simultaneous ServiceContext objects is not limited by the specification, a limitation is often forced by resource constraints.

JMF

The Java Media Framework[16] (JMF), though not part of the Java TV API, is very important for Java TV. It provides the foundation for management and control of time-based media, such as video, audio and others, in Java TV. JMF offers a player component including a GUI for playback of audio and video streams which eases the integration and flexible placement of the video presentation. Also a set of controls such as a *GainControl* for manipulating audio signal gain is provided with JMF. In the Java TV API, only controls for video size and positioning (*AWTVideoSizeControl*) and for media selection

---

[16] The Java Media Framework (JMF) - `http://java.sun.com/products/java-media/jmf/`

(*MediaSelectionControl*) have been specified, but other controls may also be implemented. A set of useful additional controls was defined in DAVIC 1.4. Also the foundation for the synchronisation of the presentation is provided by a clock mechanism of JMF.

Broadcast Data API

The Java TV API provides access to different kinds of broadcast data. Broadcast data is transmitted beside the video and audio components embedded in the television broadcast signal. The first kind of broadcast data is the broadcast file system. For transmission, broadcast carousel mechanisms are usually used. In a broadcast carousel, all files are repeatedly transmitted in a cyclic way. The data access in Java TV is modeled as the access to a conventional read-only filesystem with high access latency. Predominant protocols in the area of broadcast filesystems are the Digital Storage Media Command and Control (DSM-CC) data carousel protocol, well known from the teletext service, and the DSM-CC object carousel protocol [18]. DSM-CC is an extension of the MPEG-2 standard. The other two kinds of broadcast data are IP datagrams and streaming data. IP datagrams, unicast and multicast, are accessed via the conventional functions of the *java.net* package. Streaming data is extracted and accessed via JMF.

Application Lifecycle

Java applications for digital receivers that use the Java TV API are called Xlets. Xlets have a similar concept to Java applets. Unlike normal Java applications, Xlets have to share the JVM with other Xlets, like applets do. Therefore Xlets have a special application life cycle model and a component, the application manager, that controls and manages their life cycle. There are four states defined in the life cycle of an Xlet. Xlets are optimized for the use on TV receivers. An Xlet also has an associated context, the *XletContext*. This context is an interface between the Xlet and its environment, similarly to the *AppletContext* for applets. This interface allows the Xlet to discover information about its environment, via properties, and to inform its environment about its state changes. JavaTV provides many interesting concepts for iTV systems. Especially the application model, introduced in JavaTV, is used in all major iTV standards. The Xlet concept paved the way for interoperable iTV applications.

4.4 MHP

The Multimedia Home Platform[17] was specified by the MHP group, a subgroup of DVB. This group was created in 1997 with the goal of developing a standard for a hardware and vendor independent execution environment for

---

[17] Multimedia Home Platform - `http://www.mhp.org/`

digital applications and services in the context of DVB standards. In July 2000 the first version of the MHP standard (MHP 1.0) was published by European Telecommunications Standards Institute[18] (ETSI) [12, v1.0.3]. Only one year later MHP 1.1 became an ETSI standard [11, v1.1.1]. The latest specification is version 1.2 [7], in which support for DVB-IPTV was added. Every new version of MHP extends the previous versions.

The MHP specification defines four profiles for different classes of functionalities [24, pages 337-339]. The profiles build upon each other. MHP 1.0 specifies only the first and the second profile. The third profile was introduced with MHP 1.1 and the fourth profile with MHP 1.2. (see also [23, chapter 1]).

1. Enhanced Broadcast Profile (Profile 1): The simplest version of an MHP environment supports the Enhanced Broadcast Profile. It is aimed for set-top boxes without a return channel in a low-cost area. This profile allows the development of local interactivity applications. Because of the lack of a return channel, applications may only be downloaded from the broadcast stream - the MPEG-2 Transport Stream. Typical applications based on this profile are electronic program guides, news tickers or enhanced teletext applications.

2. Interactive Broadcast Profile (Profile 2): Additional to the functions of Profile 1 this profile includes support for a standardised return channel. Based on the return channel an interaction between the audience and the broadcast becomes possible. This enables support for applications like televoting, T-commerce or pay-per-view applications. Another advantage of the return channel is that MHP 1.1 applications can also be downloaded over an Internet connection.

3. Internet Access Profile (Profile 3): In the Internet Access Profile, Profile 2 is extended with support for Internet applications. Only APIs for accessing different Internet services and applications rather than concrete services have been specified. By using this profile, typical point-to-point services like email and WWW can be combined with the broadcast world. Online games, chat- and email-applications are often provided based on it.

4. IPTV Profile (Profile 4): The most enhanced profile is the IPTV Profile. This profile integrates support for DVB-IPTV into the MHP platform. DVB-IPTV is formed by a collection of various specifications for the delivery of DTV using IP. There are various options such as the broadband content guide (BCG) available for extending the IPTV Profile. BCG specifies signalling and delivery of TV-Anytime [13] information.

Figure 1 presents an overview of the MHP architecture. According to this figure and the presented distinction into three layers the components will be described in the following paragraphs.

---

[18] European Telecommunications Standards Institute (ETSI) - `http://www.etsi.org/`

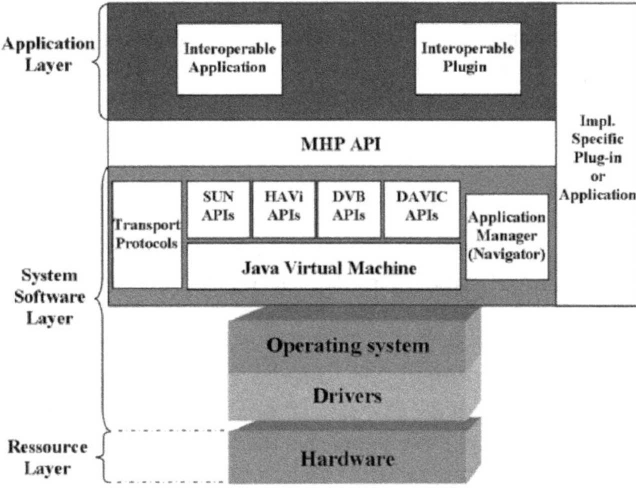

Fig. 1. MHP Architecture.

Ressource Layer

It represents the different hardware platforms of set-top boxes. Besides different components like CPU, network interface and memory, the TV specific components like the DVB frontend and the MPEG-2 decoder module also reside on this layer.

System Software Layer

Based on the hardware platform the operating system manages the integration of the hardware and offers basic functions such as process management to the MHP middleware. The first component of the middleware is represented by the Java virtual machine. The use of Java offers a hardware independent common ground for the MHP API. The Java platform of MHP is also known as DVB-Java (DVB-J) [27, pages 199-236]. PersonalJava forms the SUN API part of DVB-J. The main cause for the use of PersonalJava in MHP is the small footprint of PJVMs which fits perfectly with the limited resources of most set-top boxes. Other important Java components are the Java Media Framework (JMF), the Java TV API and the Home Audio Video Interoperability (HAVi) specification. JMF is used for controlling audio and video content. HAVi (HAVi Level 2 GUI) forms the UI components of MHP because the abstract window toolkit of Java and its UI elements were not suitable for TV UIs. Another important component of the middleware is the Navigator which forms an application manager and offers all essential functions for a user to watch TV, e.g. listing and switching channels. The application model and many APIs providing access to DTV-specific functionality of MHP stem from the Java TV specification. The applications of MHP are called DVB-J applications. The transport protocol's component in the figure consists of all parts and protocols that are necessary for communicating through different networks. TV specific protocols such as DVB-SI, DSM-CC and MPEG-2-TS and network protocols such as IP, TCP, UDP and HTTP can be found in this component.

Application Layer

As shown in figure 1, MHP is able to handle multiple applications in one JVM, on top of the MHP API. Besides the normal applications also plugins can be implemented for MHP, which are used to extend the functionality of the platform. There are two categories of applications and plugins in MHP: interoperable and non-interoperable. Interoperable applications and plugins may be used across all kinds of MHP receiver, on top of the MHP API. These DVB-J applications have been specified in Java TV. The application model of DVB-J applications follows the model of Java TV Xlets. In contrast to normal Java applications, Xlets have a similar concept to Java applets. Implementation specific applications and plugins are not interoperable. Native code or special Java APIs, which are not available in MHP, are commonly used. For that reason, such applications and plugins lose the ability of running on any kind of MHP receiver.

Besides DVB-J, there is another method for building interoperable MHP applications. The declarative language DVB-HTML for interactive TV applications has been specified since MHP version 1.1. The basic concepts of DVB-HTML such as the application life cycle had already been introduced in MHP 1.0 but has not been formalised in a specific language. The basic framework of DVB-HTML is based on a selection of XHTML 1.0 modules. For formatting, CSS level 2 is used. ECMAscript and DOM level 2 support form the basis of the dynamic aspects of DVB-HTML applications. The main reason for the introduction of DVB-HTML was that many companies had an expertise in HTML that they were willing to reuse, and that Java is not the best choice for creating presentation driven applications [23, chapter 15]. DVB-HTML is available for each of the three MHP Profiles but is most important for the Internet Access Profile.

MHP is already in use worldwide. Especially in Europe, MHP is the dominant iTV platform. Big supporters in Europe are Italy, Finland, Germany and others. Globally MHP gains importance by the fact that many other iTV specifications relate to MHP.

4.5 GEM

The Globally Executable Multimedia Home Platform (GEM) represents a subset of MHP. GEM 1.0 [10, v1.0.2] relates to MHP 1.0. It was published in the year 2003. GEM 1.1 relates to MHP 1.1 and the recent specification of GEM (1.2) [8] published in 2007 relates to MHP 1.2. The main purpose of

GEM is to enable organisations such as the CableLabs or the ATSC to define specifications based on MHP with the help of DVB. The goal is to guarantee that applications can be written in a way to be interoperable across all different GEM-based specifications and standards. GEM is not a standalone specification but a framework aimed at letting other organisations to define GEM-based specifications. GEM defines the APIs and content formats that can be used as a common basis in all interactive television standards. GEM also identifies and lists the components of MHP which are, from a technical or market perspective, specific to DVB. Other organisations are enabled to use GEM and define their own replacement of the DVB specific components as long as they are functionally equivalent. A specification in which only such DVB specific components are replaced is GEM compliant and capable of running MHP applications. Many organisations around the world have adopted GEM as the core of their middleware specification. Figure 2 shows the relationship and the functional replacements between GEM and ARIB, OCAP (cf. section 4.6) and ACAP (cf. section 4.7). GEM is referenced in its entirety in specifications like ACAP and OCAP. Differences and extensions have to be defined in detail.

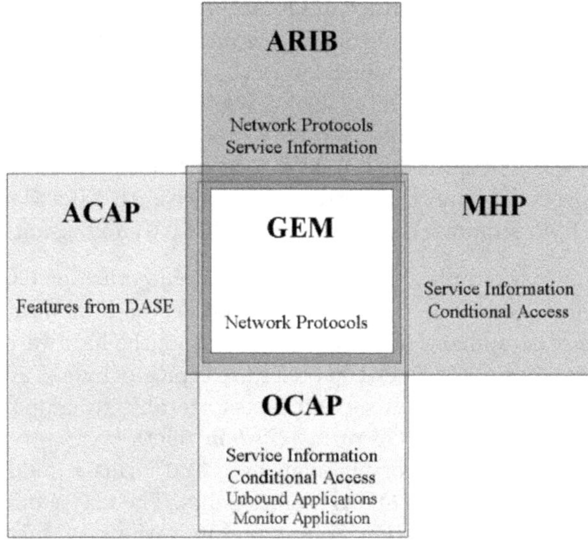

Fig. 2. Relationship between GEM and ARIB, OCAP and ACAP.

GEM seems to have the power to harmonise the iTV middleware market around the world and significantly facilitate the development of interoperable applications.

4.6 OCAP

The OpenCable Application Platform[19] (OCAP) is an open middleware standard for interactive TV. The first steps towards this standard were made by a non-profit organisation formed by many US cable television system operators called Cable Television Laboratories (CableLabs). The main goal of this initiative was to develop a middleware suitable across all set-top boxes of different vendors and all major cable TV system operators. When the work on OCAP started, its European counterpart, DVB-MHP, was still right on its way to standardisation. For that reason, DVB-MHP was investigated by the CableLabs and many parts where found to be suitable for OCAP. OCAP is largely based on MHP 1.0. Nevertheless, there are major differences between the distribution standards for digital TV used in the US and in Europe, leading to major differences in the distribution related middleware components. Also some restrictions made by the Federal Communications Commission[20] (FCC), an independent United States government agency, led to changes and extensions of the middleware components. The OCAP platform defines two profiles, OCAP 1.0 [4] and OCAP 2.0.

1. The first profile of OCAP (OCAP 1.0) was published in 2001. OCAP 1.0 defines the basic functionalities of OCAP. Over the years several versions of this profile have been published. In some versions changes were made on substantial components which led to the loss of backward compatibility between the versions. The most recent version of this profile is I16, which was released in August 2005.
2. OCAP 2.0 was first published in 2002. This profile extended OCAP 1.0 in several aspects. The most important difference with 1.0 was the inclusion of DVB-HTML support, based on the DVB-HTML extension of MHP 1.1.

In the following paragraphs, the middleware architecture of OCAP will be described. Figure 3 shows an overview of the OCAP architecture. The *Hosted Device Hardware* component of the figure represents the hardware platform of an OCAP set-top box. OCAP set-top boxes are typically hybrid analog/digital devices. This means, that such set-top boxes are able to support analog as well as digital services. The *Operating System* offers basic services such as task/process scheduling, memory management and forms a middleware layer between the hardware and the OCAP components. The major functionality of OCAP is offered by the *Execution Engine* and its various modules. Moreover, the engine provides a platform-independent interface built upon the JVM and a set of additional Java APIs. Java support in OCAP, also known as OCAP-J, is based on the DVB-J platform which was described earlier in section 4.4. We will now take a closer look at the following modules:

---

[19] OpenCable Application Platform (OCAP) - http://www.opencable.com/ocap/
[20] Federal Communications Commission (FCC) - http://www.fcc.gov/

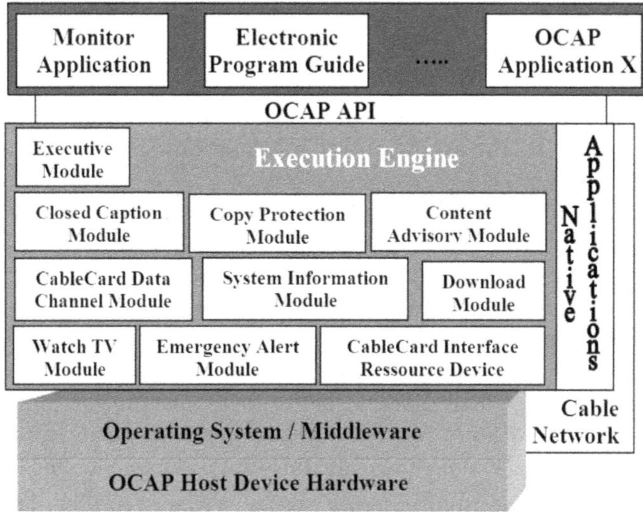

Fig. 3. OCAP Architecture.

Watch TV Module: This module offers the minimum functionality for watching TV such as switching channels. This module allows the user to watch all unencrypted channels.

Emergency Alert Module: This module is used to broadcast local or national emergency messages. Alert messages provided by the cable network operators force all receivers to show the alert message. This module is, based on the FCC rules for emergency alert system (EAS), a mandatory component of a receiver.

CableCard Interface Ressource Device: This module handles all messages of the CableCard hardware that require user-interaction, such as requesting the pin number, and satisfying the communication needs of applications via the module according to the CableCard Interface 2.0 Specification.

CableCard Data Channel Module: This module offers baseline functionality for processing data on the CableCard Data channel.

System Information Module: The System Information Module keeps track of service information. After parsing the service information, it makes the information accessible to other modules and OCAP applications. Special types of information such as emergency alert system messages are directly forwarded to the appropriate module for further processing.

Download Module: The Download Module keeps track of new available updates for the set-top box. It empowers network providers to update their set-top boxes. This is a very important function and represents the only way to get rid of erroneous firmware versions when the set-top boxes are already at the customer's home.

Closed Caption Module: The presentation of closed caption text is the main purpose of this module. It is part of the core functions and should work regardless of any extension of the network operator. It is also mandated by the FCC for all analog TV services.

Copy Protection Module: The module controls copying of analog and digital content. It controls the storage and the output of digital and analog content according to the copy control information (CCI) delivered by the conditional access (CA) system.

Content Advisory Module: The V-Chip functionality, mandated by the FCC, is handled by the Content Advisory Module. The V-Chip provides the ability to block the display of television programs based upon its rating. This offers a parental control upon the TV consumption of children. The module decodes the V-Chip signal provided via the analog broadcast and offers the rating to other modules.

Executive Module: The Executive Module is responsible for launching and controlling applications. Also the management of stored applications is handled by it. It plays a major role during the boot-up of the set-top box and starts the initial Monitor application, if one is available. If no Monitor application is available, it is responsible for controlling the receiver. While a Monitor application is running, the executive module monitors the life cycle of the Monitor application and re-launches it if it is destroyed.

OCAP applications are in many ways similar to MHP applications. The *Monitor* application, also called Monitor, plays a special role in an OCAP set-top box and represents a unique feature of OCAP. It is implemented as a single OCAP-J application or is made up of a set of applications cooperating with each other. The Monitor application has a privileged access to several APIs which are not accessible for normal applications. It helps to manage and control the life cycle of OCAP applications, cares about resource management and security issues. Monitor applications are provided by the cable television system operators and downloaded to the set-top box when it connects to the cable network for the first time. The full functionality of an OCAP set-top box is just available when the Monitor application of the network operator is present. In most cases, just the basic functions for watching DTV and using unencrypted services are available without a Monitor. By providing this application, a system operator gains much control over the set-top boxes. Further customisation can be made by the network operator because of several assumable modules of the Execution Engine such as the Watch TV Module, the Emergency Alert System Module, and so on. The Monitor application may assume the functionality of these modules. By using this feature a network operator is able to replace several functions of the box by his own implementations and create his own, branded OCAP platform. For example, the Emergency Alert System could present alert messages with a special layout or allow switching over to a channel where further information on the alert is provided[23, pages 47-52].

As mentioned before the OCAP standard has a clear focus on cable networks. For that reason OCAP is widely used by US cable TV system operators. Although OCAP is an open and mature standard a global use is unlikely because of several US-market specific characteristics.

4.7 ACAP

The Advanced Common Application Platform[21] (ACAP) is a middleware specification for iTV applications [2]. It was developed and standardised by the Advanced Television Systems Committee (ATSC), a US non-profit organisation dedicated to the development of standards for DTV. The ATSC is formed by members of the television industry ranging from broadcasters to the semiconductor industry. The ACAP standard (A/101[22]) was published in 2005. ACAP is primarily based on GEM and DTV Application Software Environment Level 1 (DASE-1) [1] but also makes use of OCAP functionalities. Many sections of the ACAP specification are references to parts of these iTV standards. An important capability of ACAP is the support of all US DTV systems, cable, satellite and terrestrial television networks. ACAP was the first attempt to harmonise the US iTV market for cable and terrestrial TV.

There are two different types of ACAP applications, procedural applications called ACAP-J and declarative applications called ACAP-X. In general ACAP-J applications are Java TV Xlets and ACAP-X, the "X" stems from XHTML, applications are very similar to DVB-HTML applications. ACAP is structured in two profiles. The first profile supports only the first application type, ACAP-J. The second profile adds support for ACAP-X applications. ACAP is based on GEM and DASE. For that reason the architecture of ACAP and its components look very similar to DASE and parts of GEM/MHP. Nevertheless there are differences between the broadcast system specific parts and parts stemming from OCAP.

4.8 Other platforms

Besides the presented open standards, several proprietary solutions for iTV middleware exist. A widespread solution is OpenTV[23]. Other products are MSTV[24] by Microsoft and MediaHighway[25] by NDS. Since these platforms are proprietary, a further description is out of scope of this paper.

5 History and Future of iTV Standards

In their latest or most advanced versions all of the presented standards feature comparable capabilities. The main differences lie in the initial objectives of

---

[21] Advanced Common Application Platform (ACAP) - http://www.acap.tv/
[22] ACAP Standard - http://www.atsc.org/standards/a101.html
[23] OpenTV - http://www.opentv.com/
[24] MSTV - http://www.microsoft.com/tv
[25] MediaHighway - http://mediahighway.nds.com/

the specifications. Goals were the development of representation languages, middleware specifications and all-round standards. Also regional distinctions led to differences and additional components. Nevertheless all the standards are strongly interrelated, as shown in figure 4. Besides several iTV standards and the HAVi specification Java plays an special role in figure 4. The application model for nearly all iTV applications is Java-based. For that reason a relation between Java and all other presented iTV standards is present. In figure 4 this relation is only indicated by arrows from Java to MHEG, DAVIC and JavaTV and the relation of these standards to all other standards. The first open standard for iTV was MHEG, initially only planned as a description language for multimedia presentations. A scripting language was included soon, in order to allow the implementation of advanced applications. In spite of other versions of the MHEG specifications, MHEG-5 and its extension MHEG-6 are the most relevant parts. These were partly included in the DAVIC specifications, the industry standard for interactive digital audiovisual applications and broadcast. Like DAVIC, MHEG-6 also makes use of the Java programming language to maximise interoperability. The JavaTV API was developed to provide a pure Java environment to control and run applications on TV receivers, such as set-top boxes. Although figure 4 does not

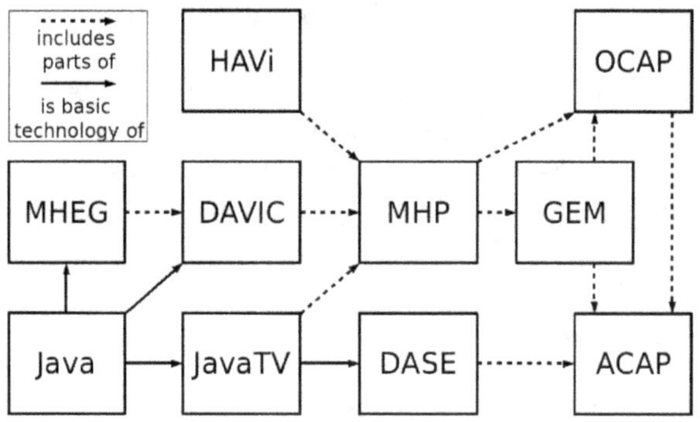

Fig. 4. Relationships between the Presented Standards.

indicate a relation between DAVIC and JavaTV, several concepts of DAVIC (e.g. controls defined in DAVIC) are used. MHP is a comprehensive specification for iTV middleware platforms. It includes parts and concepts of DAVIC, JavaTV and several parts of HAVi (e.g. UI components). Also a knowledge transfer in both directions between the standardisation of MHP and JavaTV took place. Specifically for cable networks in the US, the OCAP standard was introduced. OCAP specifies a middleware platform with several cable network and FCC specific components. The OCAP standard reuses major parts of the MHP specification. Since many standards reference MHP, a subset - GEM - was defined. GEM forms a common core for many MHP related standards. As indicated in figure 4 GEM is fully included in new versions of the OCAP specification and ACAP. ACAP is a very young standard which supports all common DTV systems in the US. Since ACAP was developed by the ATSC the former ATSC standard DASE was included in ACAP.

The emergence of the iTV standards and especially GEM shows the trend to a harmonisation of the iTV market. In the near future the slogan of Java - "Write Once, Run Anywhere" - will be true also for applications in the world of open iTV standards.

References

1. Advanced Television Systems Committee. *DTV Application Software Environment Level 1 (DASE-1) Part 1: Introduction, Architecture, And Common Facilities.* Advanced Television Systems Committee, Washington, D.C., US, 2003.
2. Advanced Television Systems Committee. *Advanced Common Application Platform (ACAP).* Advanced Television Systems Committee, Washington, D.C., US, 2005.
3. British Bradcasting Corporation. *Digital Terrestrial Television MHEG-5 Specification.* British Bradcasting Corporation, London, UK, 2003.
4. Cable Television Laboratories. *OpenCable Application Platform Specification - OCAP 1.0 Profile.* Cable Television Laboratories, Louisville, USA, 2005.
5. Bart Calder, Jon Courtney, Bill Foote, Linda Kyrnitszke, David Rivas, Chihiro Saito, James VanLoo, and Tao Ye. Java TV API Technical Overview: The Java TV API Whitepaper. Technical report, Sun Microsystems, Inc., 2000.
6. Digital Audio-Visual Council. *Statutes of The Digital Audio-Visual Council.* Digital Audio-Visual Council, Geneva, Switzerland, 1994.
7. DVB Project. *Digital Video Broadcasting (DVB); Multimedia Home Platform (MHP) Specification 1.2.* DVB Project, Geneva, Switzerland, 2007.
8. DVB Project. *GEM 1.2 (including IPTV).* DVB Project, Geneva, Switzerland, 2007.
9. European Telecommunications Standards Institute. *ETS 300 715: Terminal Equipment (TE); MHEG script interchange representation (MHEG-SIR).* European Telecommunications Standards Institute, Sophia Antipolis, France, 1994.
10. European Telecommunications Standards Institute. *ETS ES 102 819: Globally Executable MHP version 1.0.2 (GEM 1.0.2).* European Telecommunications Standards Institute, Sophia Antipolis, France, 2005.
11. European Telecommunications Standards Institute. *ETS ES 102 812: Digital Video Broadcasting (DVB); Multimedia Home Platform (MHP) Specification 1.1.1.* European Telecommunications Standards Institute, Sophia Antipolis, France, 2006.
12. European Telecommunications Standards Institute. *ETS ES 201 812: Digital Video Broadcasting (DVB); Multimedia Home Platform (MHP) Specification 1.0.3.* European Telecommunications Standards Institute, Sophia Antipolis, France, 2006.
13. European Telecommunications Standards Institute. *ETSI TS 102 822-3-1: Broadcast and On-line Services: Search, select, and rightful use of content on personal storage systems ("TV-Anytime"); Part 3: Metadata; Sub-part 1: Phase 1 - Metadata schemas.* European Telecommunications Standards Institute, Sophia Antipolis, France, 2006.

14. International Organisation for Standardisation. *ISO/IEC 13522-4:1996: Information technology – Coding of multimedia and hypermedia information – Part 4: MHEG registration procedure.* International Organisation for Standardisation, Geneva, Switzerland, 1996.

15. International Organisation for Standardisation. *ISO/IEC 13522-1:1997: Information technology – Coding of multimedia and hypermedia information – Part 1: MHEG object representation – Base notation (ASN.1).* International Organisation for Standardisation, Geneva, Switzerland, 1997.

16. International Organisation for Standardisation. *ISO/IEC 13522-3:1997: Information technology – Coding of multimedia and hypermedia information – Part 3: MHEG script interchange representation.* International Organisation for Standardisation, Geneva, Switzerland, 1997.

17. International Organisation for Standardisation. *ISO/IEC 13522-5:1997: Information technology – Coding of multimedia and hypermedia information – Part 5: Support for base-level interactive applications.* International Organisation for Standardisation, Geneva, Switzerland, 1997.

18. International Organisation for Standardisation. *ISO/IEC 13818-6:1998: Information technology – Generic coding of moving pictures and associated audio information – Part 6: Extensions for Digital Storage Media Command and Control (DSM-CC).* International Organisation for Standardisation, Geneva, Switzerland, 1997.

19. International Organisation for Standardisation. *ISO/IEC 13522-6:1998: Information technology – Coding of multimedia and hypermedia information – Part 6: Support for enhanced interactive applications.* International Organisation for Standardisation, Geneva, Switzerland, 1998.

20. International Organisation for Standardisation. *ISO/IEC 13522-7:2001: Information technology – Coding of multimedia and hypermedia information – Part 7: Interoperability and conformance testing for ISO/IEC 13522-5.* International Organisation for Standardisation, Geneva, Switzerland, 2001.

21. International Organisation for Standardisation. *ISO/IEC 13522-8:2001: Information technology – Coding of multimedia and hypermedia information – Part 8: XML notation for ISO/IEC 13522-5.* International Organisation for Standardisation, Geneva, Switzerland, 2001.

22. Thomas Meyer-Boudnik and Wolfgang Effelsberg. MHEG Explained. *IEEE MultiMedia,* 2(1):26–38, 1995.

23. Steven Morris and Anthony Smith-Chaigneau. *Interactive TV Standards.* Focal Press, Burlington, USA, 2005.

24. Ulrich Reimers. *The Family of International Standards for Digital Video Broadcasting.* Springer, Heidelberg, Germany, 2004.

25. Roger Price. MHEG: an introduction to the future international standard for hypermedia object interchange. In *Proceedings of the First ACM International Conference on Multimedia,* pages 121–128, New York, NY, USA, 1993. ACM Press.

26. Georg Ruhrmann and Joerg-Uwe Nieland. *Interaktives Fernsehen.* Westdeutscher Verlag GmbH, Wiesbaden, Germany, 1997.

27. Stephan Rupp and Gerd Siegmund. *Java in der Telekommunikation.* dpunkt.verlag, Heidelberg, Germany, 2003.

Metadata in the Audiovisual Media Production Process

Werner Bailer and Peter Schallauer

Institute of Information Systems and Information Management,
JOANNEUM RESEARCH, Graz, Austria
{werner.bailer, peter.schallauer}@joanneum.at

This chapter examines the role of metadata in the production process of audiovisual media. We analyse the role of metadata in this process and discuss strengths and weaknesses of relevant standards w.r.t. metadata involved in the process. As no single standard fulfills all requirements we present strategies for combining and mapping different metadata standards, which are crucial for interoperability in the production process.

1 Introduction

The production process of audiovisual media such as still images, audio and video consists of several stages. Following a modern view of this process, the term production does not just refer to the act of creation of an audiovisual media item, but encompasses many stages in its lifecycle. In the pre-production a media item is designed and conceptualised, then material is actually captured. In the post-production material from various sources and of different types is combined into a new media item, which is then archived, annotating it in order to make it accessible for consumption or reuse and delivered and presented to an audience. Of course the boundaries between these stages are fuzzy, for example, in the production of entirely computer generated movies production and post-production blend into one another and in the case of interactive media the content gets its final shape during presentation.

We are mainly considering the professional production process here, however, many aspects also apply to content generated by home or amateur users. One of the most important domains where this process is applied is broadcasting, including for example the production of news, documentaries and TV series and the management of large audiovisual archives. Movie production is of course a related area, but it is different due to the fact that it is more project oriented, with very few links between different productions and practically no re-use of content. In contrast, news production is an area where material re-use is very common: parts of a newscasts are used again and again while a

W. Bailer and P. Schallauer: *Metadata in the Audiovisual Media Production Process*, Studies in Computational Intelligence (SCI) **101**, 65–84 (2008)
www.springerlink.com

story is developing, then in magazines that add some background information and even months or years later when a related story is in the news. In addition, content is sold to other broadcasters all over the world who integrate it into their productions. As the audiovisual media production process is a field centered around audiovisual content, metadata describing this content plays of course a crucial role.

In the remainder of this section we discuss the stages of the production process in more detail. Section 2 analyses the different types of metadata as well as their properties and roles in this process. After this analysis, Section 3 starts with describing a set of requirements for a metadata model to be used in the production process. After reviewing existing metadata standards and formats we discuss them in the light of these requirements. We will see that there is no single standard that fulfills all requirements and thus dealing with different standards is necessary in practice. Section 4 treats this topic, discussing the integration of different standards and formats, conversion between them as well as editing metadata descriptions. Section 5 concludes this contribution.

First of all, it is necessary to define some of the terms we are dealing with throughout this contribution, following mainly the conventions used by the Society of Motion Picture and Television Engineers (SMPTE). The audiovisual material itself – of whatever type – is called an *essence*. *Content* is defined as essence plus the *metadata* describing the essence. Together with the rights to use it, the content becomes commercially relevant and is thus called *asset*. By *media item*, we denote a single unit of content. It must be noted that one unit of content (e.g. a movie) is not necessarily related to one physical unit of *media* (e.g. a reel of film, a tape). In practice, there are media holding multiple contents and contents stored on multiple media. The relation between content and the media it is stored on is thus part of the metadata description of the content. We use the term *segment* following the MPEG-7 standard [22] as a piece of content created by decomposing the original content by an arbitrary criterion (e.g. shots over time, regions of the same colour in an image).

Figure 1 shows an overview of the production process[1]. As noted above, the borders between the stages are becoming less clear. For simplicity we nonetheless describe the classic stages of the process. The *pre-production* stage encompasses everything from the first ideas to the actual start of gathering material for the production. While no essence is created in this stage, a lot of metadata is produced, for example a script, sketches of the scenes, photographs of shooting locations[2], etc. The *production* stage includes the creation of the essence, traditionally shooting film or video material, but also doing animation, sound recording, etc. While the main result of production is essence a

---

[1] A description of the video production process based on the canonical process model of media production currently under development can be found in [27].

[2] Note that although they are essence themselves, they are metadata related to the content being produced.

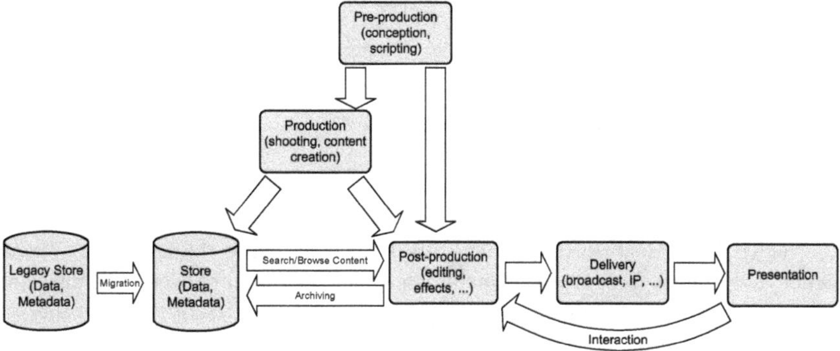

Fig. 1. An overview of the audiovisual media production process. Apart from newly produced essence and metadata also the migration of legacy content is an issue (cf. Section 2.1).

lot of metadata is also created, such as the date and time of shooting, measurements of the set that are needed to insert special effects later, and so on. Due to the growing amount of artificially generated instead of shot content, production merges more and more with the subsequent *post-production* stage of the process. In classical post-production, the main task is editing the shot material, possibly combining it with archive material. Other tasks are insertion of special effects and titles, as well as colour correction. This stage rather consumes metadata, the most important metadata created in that stage is the description of the process, such as the editing operations performed (the so-called *edit decisions*), the colour grading operations, etc. The outcome of the post-production stage is mainly the final edited essence, with usually very few metadata (mainly global metadata describing the content as a whole, e.g. title, some production information, an abstract, and in some cases some more detailed metadata such as captions). Most of the metadata stays in the production environment and is not available later. This causes a major problem in *archiving*, where metadata must be recreated in order to make content searchable, a process called "documentation". Due to the high costs related to recreating the metadata, most content is only documented by global metadata, only content from which clips may be commercially relevant later (such as news stories) are described in more detail. The *delivery and presentation* of content to an audience is typically the stage where both essence and metadata are consumed. However, in new media such as interactive television the result of the post-production stage is not a final product, but an intermediate result, and the final selection and ordering of the content is done during presentation, based on the users interactions. This also creates metadata in the form of new possible edit decisions and user feedback.

Source	Effort	Abstraction	Confidence
Capture	–	–	+
Legacy and related information	o	+	o
Manual annotation	+	+	+
Content analysis (low- and mid-level)	–	–	o
Content analysis (high-level)	–	+	–

Table 1. Comparison of the properties of metadata sources (– ... low, o ... moderate, + ... high). Effort denotes the human work needed to create the metadata. The abstraction level ranges from low-level metadata, which is directly derived from the signal, to high-level semantically meaningful metadata. Confidence describes the fidelity and reliability of the metadata source.

2 Analysis of Metadata in the Process

A basic classification of the types of sources can be easily made, although for some of the classes, there exists a great variety. A classification of metadata by type is difficult, as many different criteria can be used.

2.1 Sources of Metadata

Digital essence originates either from a digital capture device ("digitally born content") or is the result of digitisation (e.g. scanning film). When looking at the source of the metadata for the digital content, we can discriminate the categories described below. Table 1 shows the required effort for metadata creation, the abstraction level and the confidence of the different sources.

A basic distinction can be made between so-called *technical* metadata describing basic properties like sampling and encoding, which are necessary to interpret the essence correctly, and *descriptive* metadata describing content in a more comprehensive way (we will discuss this in more detail in Section 3.3). Capture is primarily a source for technical metadata, while the other sources provide mainly descriptive metadata.

Capture

Like the essence itself, metadata can be captured during the creation of digital content, using either the same or other capture devices. An important group of metadata are the settings of the device used for capturing the essence, such as the Exif information [14] describing exposure time, resolution, focal length, etc. of digital still image cameras. In professional and increasingly also in consumer cameras the raw sensor data are stored, together with a lookup table (LUT) that maps the sensor data to a standard colour space. Other commonly captured metadata are date, time and place, either absolutely (e.g. using GPS modules) or relatively (e.g. camera cranes logging their movements). In the case that analog material is digitised, information about quality and defects in the material can be captured.

Legacy and Related Information

In many cases metadata is created before the essence is even created (e.g. a movie script) or at least digitised (such as archive catalogues on paper or in proprietary databases, labels on tapes and film cans). A similar case is metadata that is not directly related to the production process, but none the less valuable for the description of the content, such as works based on the content (e.g. summaries and reviews) or related to the content (e.g. a newspaper article about the same event as a story in a news broadcast, a book on the same topic as a documentary). All this information has to be migrated in some way (e.g. digitised) to be usable.

Manual Annotation

Manual annotation of content is a reliable and valuable source of metadata, as it can provide semantically meaningful abstraction. However, in most application areas it is prohibitive due to the high effort (i.e. several hours per hour of video) and thus cost. As a compromise, manual annotation is often only done on a per content basis, describing its global properties.

Content Analysis

Another source of metadata is the automatic analysis of the content in order to extract metadata describing it. There are several levels of content analysis, starting from the extraction of parameters from the visual or audio signals up to the analysis of the high-level structure of content and events. In most cases the higher level tools depend on the lower level tools. The problem of extracting semantics from the low- and mid-level features is known as the semantic gap [32] and is not satisfactorily solved. However, the use of domain knowledge can alleviate it. The advantage is that automatic content analysis allows to process large amounts of essence at low cost, with the drawback of yielding mainly low- and mid-level feature descriptions at a limited confidence. More detailed information about content analysis tools and extensive references can be found in [17, 35, 3].

Text and Semantic Analysis

Text is still the most commonly used modality for the extraction of semantic information. This includes recognition of references to named entities (e.g. persons, organisations, places) as well as linking them to ontological entities, the detection of topics and the classification of content segments and linking content to legacy or related information.

2.2 Metadata Properties

There are a number of criteria by which metadata can be classified. In the following we discuss a set of properties that can serve to distinguish different kinds of metadata.

Scope. Metadata elements may refer to the whole content (global metadata) or to just a segment of it[3]. They can apply to a spatial, temporal or spatiotemporal segment of the content. The same metadata elements may exist in different scopes, such as the title of a movie and the title of a scene.

Data type. Metadata elements may also have different data types. They can be either textual or numerical, each with either continuous (free text in the case of text) or a discrete set of values (controlled text). A special case of controlled text are vocabularies like thesauri or ontologies, where the reference to a term is just its identifier. In the case of low-level descriptors, the data types may be a vector or matrix.

Time dependency. The value of a metadata element can be constant over a whole segment or change over time. In [16], these types of metadata are called *static* and *dynamic* metadata respectively.

Spatial dependency. Similar to the temporal dependency, metadata elements may refer to all of the content or just to a spatial/spatiotemporal segment (e.g. a static or moving object).

Modality/channel dependency. A metadata element may also refer to all modalities of a content (e.g. genre is sports), to a specific modality (e.g. colour distribution of a shot) or even just one channel (e.g. spoken text in the center audio channel). Whether metadata refers to all modalities or just a specific one strongly depends on the type of metadata. From its nature, low-level metadata is usually specific to one modality, while high-level metadata (textual and semantic metadata) is often related to all modalities of the audiovisual content.

Context dependency. Context dependency mainly concerns high-level metadata. The metadata describing a segment of content can only be valid in a certain editorial context. This is especially true for so called affective content description, i.e. annotation describing the emotions evoked in the audience (e.g. classifying a segment as "frightening").

2.3 The Role of Metadata in the Production Process

The pre-production stage is the step of the workflow where the awareness for the importance of metadata is limited, although a lot of valuable high-level metadata is produced in this stage. Thus interoperability with later steps is often not possible.

---

[3] In the audiovisual archive domain this is called *synthetic* and *analytic* documentation respectively.

In the production stage metadata is captured together with the essence. Due to the fact that the workflow is increasingly becoming fully digital there is growing interest in the metadata that can be gathered during production. An important example are camera parameters (e.g. motion of the camera crane, lens settings) and measurements of the scene in order to use them for later insertion of special effects in 3D space. This is often needed in virtual studio applications for broadcasting.

The post-production workflow is a very heterogeneous one, involving a number of different tools and systems. Metadata is important to enable interoperability between these systems (which is in the interest of the user, though not always of the manufacturers). As soon as non-linear editing systems have been used, simple de-facto standards such as the text-based edit decision lists (EDLs) have emerged. With computers becoming fast enough to do colour correction in software and thus makes it an ubiquitous task, similar standards have been defined for colour grading information [7]. Recently the industry has started to define standards for exchanging essence packaged with metadata, e.g. DPX [13] and MXF [26], as well as more expressive formats for editing information [1].

In audiovisual archives metadata is crucial as they are the key to using the archive. The total holdings of European audiovisual archives have been estimated to 10^7 hours of film and 2×10^7 hours of video and audio respectively [41]. More than half of the audiovisual archives spend more than 40% of their bugdet on cataloguing and documentation of their holdings, with broadcast archives spending more and film archives spending less [11]. Large broadcast archives document up to 50% of their content in a detailed way (i.e. describing also the content over time), while smaller archives generally document their content globally. However, all of them mainly document finished and published items, but hardly raw material and other edited items. A major problem is still that mosts archives use their own metadata and documentation models. Initiatives such as the EBU Metadata Exchange Scheme P_Meta ([30], see Section 3) try to overcome this problem by specifying standard exchange formats between archives, content creators and distributors.

In the delivery and presentation of content, metadata is of growing importance. With traditional systems such as analog TV the situation has been quite simple: Each of the common systems (e.g. PAL, NTSC) has fixed number of lines, frame rate and primary colours. Similarly, 35mm film is a well-defined standard. With the convergence between different media and growing variability of presentation devices metadata becomes more relevant, both in the professional and consumer area. In digital cinema projection different resolutions, frame rates and colour models are supported in order to enable cinema owners to display not only movies following the specification of the major Hollywood studios [9], but also independent movies and other content, such as HDTV sports broadcasts (e.g. games of the soccer world cup 2006 were commercially successful in cinemas[4]). Management of content is an issue in digital cinema playout, handling versions in different languages, supporting different sound systems, 2D and 3D versions, etc. In the consumer area, adaptation of content to a large variety of end user devices, from TV screens to mobile devices is an important topic. There is specific metadata for content adaptation, such as those defined in MPEG-21 [21].

3 Metadata Modelling

After analysis of the different sources and types of metadata as well as their properties and roles in the production process the aim of this section is to outline a metadata model covering all metadata of the production process. After describing a set of requirements for the model, existing metadata standards are reviewed and discussed in the light of these requirements.

3.1 Requirements

Based on the process discussed above the following list of requirements for a metadata model can be stated.

Comprehensiveness. The metadata model must be capable of modelling a broad range of multimedia descriptions (e.g. descriptions of different kinds of modalities and descriptions created with different analysis and annotation tools), covering the variability in terms of types, properties and sources discussed in Section 2.

Fine grained representation. The metadata model must allow describing arbitrary fragments of media items. The scope of a description may vary from whole content to small spatial, temporal or spatiotemporal fragments of the content.

Structured representation. The metadata model must be able to hierarchically structure descriptions with different scopes and descriptions assigned to fragments of different granularity.

Modularity. The metadata model should not introduce interdependencies within the description that do not exist in the content, such as between content analysis results from different modalities (e.g. speech segments and visual shots). The metadata model shall also separate descriptions with different levels of abstraction (e.g. low-level feature descriptions and semantic descriptions). This is important, as descriptions on higher abstraction levels are usually based on multiple modalities and often use domain specific prior knowledge, while low-level descriptions are typically specific to one modality.

---

[4] see   http://www.hollywoodreporter.com/hr/content_display/
international/news/e3if9a2ba713cf749a967183b476a76660a

Extensibility. It must be possible to easily extend the metadata model to support types of metadata not foreseen at design time or which are domain or application specific.

Interoperability. It shall be easily possible to import metadata descriptions from other systems or to export to other systems.

3.2 Standards and Formats

A large number of standards for representing audiovisual metadata exist. They come from different organisations and cover diverse application areas. Practically all multimedia file formats include technical metadata of the content, for example Exif [14] metadata in still image formats. This is especially true for container formats that combine essence and associated metadata in one file. The Material Exchange Format (MXF) [26] is an example for such a container format (discussed in detail below). The Digital Picture Exchange (DPX) [13] format is commonly used for still image sequences in digital cinema production and also allows to store a vast range of metadata in its header, supporting the same metadata format as MXF. Another example for a container format is the Digital Cinema Package (DCP) [9], used to transport digital movies and associated metadata to cinemas. The following overview does not discuss all the formats capable of holding some technical metadata but includes only standards that are relevant in audiovisual media production and that offer at least basic support for descriptive metadata.

Dublin Core

The Dublin Core metadata standard [10] was originally developed to describe electronic text documents but has later been extended to also cover audiovisual material. Focusing on simplicity it contains fifteen elements belonging to three groups (content, version and intellectual property). Some of these elements can be refined using qualifiers to narrow down the semantics ("Qualified Dublin Core"). The content of the elements is primarily text without further inner structure. Dublin Core descriptions are represented using XML.

MPEG-7

The ISO/IEC standard *Multimedia Content Description Interface* (MPEG-7) [22] has been defined as a format for the description of multimedia content in a wide range of applications. MPEG-7 defines a set of description tools, called description schemes (DS) and descriptors (D). Descriptors represent single properties of the content description, while description schemes are containers for descriptors and other description schemes. The definition of description schemes and descriptors uses the *Description Definition Language (DDL)*, which is an extension of XML Schema. MPEG-7 descriptions can be either represented as XML (textual format, TeM) or in a binary format (binary format, BiM).

A core part of MPEG-7 are the *Multimedia Description Schemes (MDS)*, which provide support for the description of media information, creation and production information, content structure, usage of content, semantics, navigation and access, content organisation and user interaction. Especially the structuring tools are very flexible and allow the description of content on different levels of granularity. In addition, the *Audio* and *Visual* parts define low- and mid-level descriptors for these modalities.

The concept of profiles has been introduced to define subsets of the comprehensive standard which target certain application areas. Three profiles have been standardised: the *Simple Metadata Profile (SMP)*, which describes single instances or collections of multimedia content, the *User Description Profile (UDP)*, containing tools for describing personal preferences and usage patterns of users of multimedia content in order to enable automatic discovery, selection, personalisation and recommendation of multimedia content, and the *Core Description Profile (CDP)*, which consists of tools for describing general multimedia content such as images, videos, audio and collections thereof.

EBU P_Meta

The European Broadcasting Union (EBU) has defined P_Meta [30] as a metadata vocabulary for programme exchange in the professional broadcast industry. It is not intended as an internal representation of a broadcaster's system but as an exchange format for programme-related information in a business-to-business use case. P_Meta consists of a number of attributes (some of them with a controlled list of values), which are organised in sets. The standard covers the following types of metadata: identification, technical metadata, programme description and classification, creation and production information, rights and contract information and publication information.

Material Exchange Format

The Material Exchange Format (MXF) [26] is a standard issued by Society of Motion Picture and Television Engineers (SMPTE), defining the specification of a file format for the wrapping and transport of essence and metadata in a single container. The Material Exchange Format is an open binary file format targeted at the interchange of captured, ingested, finished or "almost finished" audio-visual material with associated data and metadata. Support for technical metadata is built directly into MXF specification. In order to provide enough flexibility to deal with different kinds of descriptive metadata, a plugin mechanism for descriptive metadata is defined. These descriptive metadata schemes (DMS) can be integrated into MXF files. So far SMPTE has standardised the Descriptive Metadata Scheme 1 (DMS-1) and the EBU has defined a DMS for P_Meta.

DMS-1 The SMPTE Descriptive Metadata Scheme 1 (DMS-1, formerly know as Geneva Scheme) [12] uses metadata sets defined in the SMPTE Metadata Dictionary (see below). Metadata sets are organised in descriptive metadata (DM) frameworks. DMS-1 defines three DM frameworks, that correspond to different granularities of description: production (entire media item), clip (continuous AV essence part) and scene (narratively or dramatically coherent unit). When DMS-1 descriptions are embedded into MXF files they are represented in KLV[5] format, but there exists also a serialised format based on XML Schema.

SMPTE Metadata Dictionary The SMPTE Metadata Dictionary [34] is not a metadata format on its own, but a large thematically structured list of narrowly defined metadata elements, defined by a key, the size of the value and its semantics. It is used for all metadata embedded in MXF files, but the elements defined in the dictionary are also used outside MXF.

BBC Standard Media Exchange Framework (SMEF)

SMEF [8] is a data model defined by the BBC to describe the metadata related to media items (media objects) and programmes and parts thereof (editorial objects), down to the shot level. In contrast to P_Meta, it was primarily designed for internal use and not as an exchange format. Thus SMEF is expressed as a data dictionary and a set of entity relationship diagrams, but no serialised representation is defined.

Controlled Vocabulary and Ontologies

Audiovisual content descriptions often contain references to semantic entities such as objects, events, states, places, and times. In order to ensure consistent descriptions (e.g. make sure that persons are always referenced with the same name) controlled vocabulary should be used. The simplest form of a controlled vocabulary is a list of possible values of a property (e.g. the ISO 3166 list of countries). Taxonomies are defined as a tree structure of terms and simple relations between them, and allow the term names and definitions to be multilingual. Thesauri are hierarchical (or poly-hierarchical) structures of terms in a given application domain. In a thesaurus more complex relations such as the hierarchy of terms, synonyms, related terms (or near-synonyms) can be described, both mono- and multilingually. In museums, archives, and libraries *Authority Control Files* [19] are used to describe entities such as persons, corporate bodies, and families. It enables annotators to differentiate items with similar or identical terms by adding additional information to the records. The information managed with controlled vocabularies can also be represented using ontologies (and standards such as RDF/OWL).

3.3 A metadata "taxonomy"

Apart from organising metadata by its properties and sources (as we have done in Section 2) it seems to be useful to classify them by type and functionality.

---

[5] KLV (Key-Length-Value) encodes items into Key-Length-Value triplets, where key identifies the data or metadata item, length specifies the length of the data block, and value is the data itself [33].

	DC	MPEG-7	P_Meta	DMS-1	SMPTE MD
Identification	+	++	++	+	+
Production	o	+	++	o	+
Rights	o	o	++	o	+
Publication	o	o	++	o	+
Process related	−	−	−	o	+
Content related	−	++	o	+	o
Relational/enrichment	o	+	o	o	+

Table 2. Coverage of metadata classes by different standards (− ... not supported, o ... basic support, + ... good support, ++ ... comprehensive support).

In fact, most metadata standards classify the supported elements by these criteria, however, the classifications are not congruent. In the following, we use a categorisation that is based on the metadata standards described above.

Technical metadata are widely supported by a variety of standards. The SMPTE Material Exchange Format (MXF) standard [26] calls technical metadata needed to understand the organisation of the essence in the file *structural* metadata. The SMPTE Metadata Dictionary [34] defines many additional technical metadata elements. In MPEG-7 [22], the *MediaInformation DS* contains both many technical metadata elements, but also media identifiers and media locators (pointers to physical or electronic instances of the essence). In P_Meta [30], the technical metadata sets also include the media locators, while in the SMPTE Metadata Dictionary [34] both identifiers and locators are a separate group. A specific kind of technical metadata are those related to the quality and defects of the essence. In the SMPTE Metadata Dictionary, there are quality related metadata both in the parametric and in the process metadata groups. MPEG-7 provides basic support for quality description in the *MediaInformation DS*, while the more detailed *AudioSignalQuality DS*, which supports further quality parameters and description of defect events), is a audio descriptor added in [23].

The most commonly used descriptive metadata elements can be classified as follows (Table 2 provides an overview of the support of metadata classes in the different standards).

Identification. Identification information contains usually IDs as well as the titles related to the content (working titles, titles used for publishing, etc.). In some formats identification metadata constitutes a separate group of metadata, in MPEG-7 this information is part of the *CreationInformation DS*.

Production. This describes metadata related to the creation of the content, such as location and time of capture as well as the persons and organisations contributing to the production. In MPEG-7 this information is part of the *CreationInformation DS*, in P_Meta production metadata complements identification metadata and the SMPTE Metadata Dictionary contains a group called *administration*, which contains production and rights metadata.

Rights. In the business-to-business oriented P_Meta standard, rights related information is a separate large subgroup, while in MPEG-7 this information is part of the *CreationInformation DS*.

Publication. Publication information describes previous use (e.g. broadcasting) of the content and related information (e.g. contracts, revenues). In P_Meta production metadata complements identification metadata, in MPEG-7 this information is contained in *UsageDescription DS*.

Process-related. This is a separate group in the SMPTE Metadata Dictionary, describing the production and post-production history of the essence, e.g. information about capture, digitisation, encoding and editing steps in the workflow.

Content-related. Content-related metadata is descriptive metadata in the narrowest sense. An important part is the description of the structure of the content (e.g. shots, scenes). MPEG-7 provides comprehensive structure description tools for that purpose. In MXF DMS-1 structuring is realised by so called frameworks, related to the complete content, clips and scenes. Content related metadata also includes the textual and semantic description of the content, keywords, segment classification, etc.

Relational/enrichment information. This information describes links between the content and external data sources, such as other multimedia content or related textual sources. In MPEG-7 this information is part of the *CreationInformation DS*, in P_Meta it is ancillary information to identification and the SMPTE Metadata Dictionary contains a *relational* metadata group.

3.4 Discussion

The diversity of the investigated standards is high. None of the standards supports all types of metadata stemming from the production process in a comprehensive way (see Table 2) and fulfills all the requirements to a metadata model (see Table 3). Technical and identification metadadata are supported in all standards. P_Meta supports very well metadata necessary for business-to-business exchange of programmes (production, rights and publication metadata). The SMPTE Metadata Dictionary provides at least basic support for metadata of the entire production process, but does not contain any structuring capabilities. Only MPEG-7 and DMS-1 support description of content structure well. The support for describing content semantics, e.g. the occurence of objects within a video and their actions and relations, is only provided by MPEG-7, however in a limited way. The main restriction is the lack of formal semantics (cf. below). Thus inference of new knowledge out of the semantic metadata (reasoning) is not fully possible. For reasoning appropriate metadata representations need to be utilised, e.g. RDF/OWL.

A problem of nearly all the standards discussed above is the deficit of well-defined semantics of their description elements. This is very much related to the breadth of the intended application area. For example, P_Meta is defined for quite specific use cases, thus this problem is not so severe there, as the semantics of the elements is defined more precisely. On the other end of the scale is Dublin Core, where the semantics of the elements are very fuzzy, e.g. the element `dc:creator` could be used by one user to annotate the production company, while another may use it for the author or cinematographer. Although most of the technical metadata in the SMPTE Metadata Dictionary has very clear semantics, a number of similar cases exist. This lack of semantics causes interoperability problems when exchanging documents conforming to one standard between different organisations or systems.

The specification documents of many standards contain some textual descriptions of the semantics of the elements being defined. For all standards targeted at a broad application area, the semantics of some elements remain necessarily fuzzy. In contrast to other standards, MPEG-7 provides at least a partial solution for this problem. As mentioned in Section 3.2, the concept of profiles has been introduced in MPEG-7 to define subsets targeted at specific application areas. As outlined in [24], one of the part of the definition of a profile is a set of constraints of the semantics of the description schemes and descriptors in the profile, which can be much more restrictive when only the target application area of the profile is considered. However, the profiles adopted in the standard [25] do not contain definitions of semantic constraints. The Detailed Audiovisual Profile (DAVP) [4] has been proposed for covering many of the requirements of the audiovisual media production process and also contains a set of semantic constraints on the included metadata elements, mainly concerning structuring of the description.

However, whether the description of semantics is contained in the specification of the standard (such as for many P_Meta elements) or added in a profile definition (such as in MPEG-7 DAVP), one problem remains: it is still in textual form. This means that it is not possible to use it for validating documents in terms of semantics to a certain standard or for mapping elements between different standards. This shortcoming of MPEG-7 and other standards has been criticised in several works [38, 29, 39] and has led some researchers to discard traditional metadata standards and use metadata models based on Semantic Web technologies such as RDF/OWL for describing multimedia content [2, 15, 18, 40]. As discussed in Section 4.1 there are types of metadata for which these representations are very suitable, but they are not a full replacement for the audiovisual metadata standards that have emerged.

An alternative approach is to complement the standard definition by a formalisation of the semantics of the description elements. This approach (using OWL and rules) has been outline in [38] for the MPEG-7 DAVP profile. This has led to the implementation of a service, that is capable of validating documents w.r.t. the semantics of the profile [37]. The latter work also outlines how this approach could be applied to other metadata standards.

	DC	MPEG-7	P_Meta	DMS-1	SMPTE MD	SMEF
Comprehensiveness	−	+	o	o	+	o
Fine grained representation	−	+	−	+	+	−
Structured representation	−	+	−	+	−	−
Modularity	−	o	+	+	+	+
Extensibility	−	+	−	+	+	−
Interoperability	−	−	+	−	o	o

Table 3. Requirements on a metadata model for the audiovisual media production process and fulfillment by different standards (− ... low, o ... moderate, + ... high).

4 Coping with Diversity

As we have seen from the discussion above, there is no single metadata format or standard that optimally fulfills the diverse requirements across the production process. In practical applications one thus has to deal with combining different metadata formats. Conversion and mapping between between different formats is crucial for interoperability across the production chain. Another aspect is the loss of metadata due to editing, which motivates research into automatically applying edit decisions to metadata descriptions.

4.1 Heterogenous Metadata Models

As we have seen, the existing metadata standards have different strengths and focus in terms of the properties and types discussed in Section 2. Even if it may be possible to represent a metadata element in a certain format it may not be efficient due to the type of representation used.

The consequence of this fact is to use different formats and standards for different kinds of metadata, selecting the most appropriate one for each. However, it is often necessary – and indeed beneficial – to jointly use different kinds of metadata. For example, high-level audiovisual analysis such as scene segmentation or event detection can benefit from using all available modalities, from low-level visual features to semantic analysis of the spoken text, from audio classification to related articles on the Web. It is thus necessary to create metadata models that are heterogeneous, be it in terms of the application scope of the metadata elements or their abstraction level.

Integrating Different Application Scopes

As the different multimedia metadata standards and formats have emerged from different communities, they focus on different application areas. Metadata specific to one area cannot or only difficultly be represented in other formats. Examples for such metadata elements are the publication-related metadata in P_Meta, which is very comprehensive as the standard is targeted toward professional broadcasters, or the process-related metadata in the SMPTE Metadata Dictionary.

In practical applications this means finding ways of combining different formats. This task is facilitated by the fact that there are two common representations, which are both designed to support extensibility: one is XML and the other is KLV. Both representations are designed to allow an application to skip unknown parts of a document. However, the standards do not always permit the use of this flexibility, for example, P_Meta does not provide an extension mechanism. In contrast, the MPEG-7 conformance rules (part 7 of [22]) allow both extension from MPEG-7 elements as well as embedding MPEG-7 elements in other documents. Also MXF foresees hooks to include metadata in other formats ("dark metadata").

An example for an approach combining different standards is the system for broadcast archive documentation presented in [5], which uses subsets of MPEG-7 (for the description of the content and its structure) and of P_Meta (for production and publication information). Another example is the TV Anytime standard[6], which defines a set of own content description metadata elements, but uses the MPEG-7 user description profile (UDP) for the representation of user preferences.

Integrating Different Abstraction Levels

Another reason for the need to integrate between different metadata representations is the fact that they focus on specific abstraction levels. Standards like the SMPTE Metadata Dictionary and especially MPEG-7 provide tools for describing low-level signal related parameters, P_Meta and Dublin Core for example mainly focus on textual information, and on the other end of the scale semantic representations such as RDF provide the highest semantic expressivity.

It is hardly possible to decide for just one of the representations. It has been reported that MPEG-7 does not well support formal semantics of the descriptions [18, 28, 36] (and the conclusions are applicable to similar formats), which has led to a number of attempts to model the MPEG-7 description tools using RDF (e.g. [15, 40]). Representations like RDF are well suited for high-level information. However, trying to represent low- and mid-level features results in RDF descriptions that are inefficient for operations such as similarity matching. Often these documents also contain numbers of triples that might go beyond the current capacity limits of RDF stores (consider for example the description of visual descriptors of the key frames of several hours of video) [37]. Another aspect is that the description of audiovisual content is related to one audiovisual essence, while ontological knowledge such as named entities and their relations are usually valid at least for a collection of content (e.g. one archive).

A possible solution is thus to store ontological information centrally at one site, while describing multimedia content per document. This approach is for

---

example used in the system for broadcast archive documentation presented in [5]. The data model is the hub for integrating the different representations. A central knowledge base stores the named entities and categories for an archive (in this case the KIM platform [31] is used as knowledge base). The entities are linked from the audiovisual description using the MPEG-7 semantics description tools, which are flexible enough to not only use MPEG-7 classification schemes but any concept from external controlled vocabulary, such as the knowledge base.

4.2 Mapping, Conversion and Editing

The audiovisual media production workflow is a heterogeneous one, involving a number of different systems and applications. Nearly all of them deal with some kind of metadata and many of them support specific formats. The same type of metadata may be represented differently in different stages of the workflow. For example, a remotely controlled camera crane has a specific format for describing the movements it performs. In post-production, the same information is useful for inserting special effects, however, the 3D modelling software might need another format for describing the trajectory of its virtual camera than the camera crane.

Mapping and conversion between different metadata representations is thus an inevitable step. The complexity of the problem depends mainly on two aspects: well-definedness of metadata elements and incongruence between formats.

As discussed above, there are many types of metadata, some of them describing precisely defined low-level features, while others describe semantic and affective properties of the content, and may have rather fuzzy definitions. For example, the text of the Dublin Core field "description" may convey different semantics depending on which person (from which organisation) created it.

For some metadata elements, there exist clear one-to-one mappings between different metadata standards. But for many elements, the semantics of the metadata elements in different formats are only similar or overlap. The meaning of the metadata element in different formats may be different, undefined or just informally defined (i.e. by the way how it is normally used). For example, the element "title" existing in many formats may contain the main or subtitle, a working title, that of a translation, etc.

As a consequence, mapping between different standards can only be done if the semantics of the element are clearly specified or encoded into the mapping application. If mapping or conversion shall be done automatically, formalisation of semantics is required (cf. [38] for a discussion of this problem for mapping between different MPEG-7 profiles).

A related problem is metadata editing, i.e. the creation of metadata for the new content (the result of the editing process) using the metadata that may be available for the different source materials. In today's media production processes, the outcome is the edited essence. A large part of the metadata that has been available throughout the production process is not available any more at the end of the chain, as it is lost during editing.

An approach for metadata editing using a proprietary XML-based metadata format is presented in [20]. A framework for metadata editing is proposed in [6], with an example implementation working on MPEG-7 documents. The framework solves some of the merging issues by using formalisation of the semantics of the metadata elements.

5 Conclusion

In this contribution, we have analysed the production process of audiovisual media and the role of metadata in this process. This process involves many stages and combines them in a heterogeneous way, thus involving a number of different tools and applications. We have seen that the metadata involved in the process stems from various sources and differs in its properties and abstraction level. Practical applications have to deal with this diversity. The more dynamic the production workflow becomes, the more important is the role of metadata for establishing interoperability.

We have stated a set of requirements for a metadata model that is capable of covering all stages of the production workflow. After reviewing the existing standards in the light of these requirements, it becomes clear that – although there exist very comprehensive and relevant ones – no single standard or format can fulfill all the requirements. It is thus inevitable to deal with combination and conversion of these standards and formats. We have grouped the metadata elements by their types and properties, which is not only a theoretical exercise, but also given valuable hints about how to organise the metadata elements involved in the data model of a concrete application.

A drawback in today's media production process is that there is still a lack of interoperability in terms of metadata between the applications and systems involved. This is the main obstacle for using the full potential a common metadata workflow can offer to increase interoperability and efficiency in the production process. The formalisation of semantics of metadata standards or profiles (cf. Section 3.4) and the approaches for integrating heterogeneous metadata models, mapping between different formats and metadata editing presented in Section 4 contribute to solve these interoperability problems.

References

1. Advanced Authoring Format (AAF) Edit Protocol. AAF Association, 2005.
2. Richard Arndt, Raphal Troncy, Steffen Staab, and Lynda Hardman. Adding Formal Semantics to MPEG-7: Designing a Well-Founded Multimedia Ontology for the Web. Technical Report KU-N0407, University of Koblenz-Landau, 2007.

3. Werner Bailer, Franz Hller, Alberto Messina, Daniele Airola, Peter Schallauer, and Michael Hausenblas. State of the Art of Content Analysis Tools for Video, Audio and Speech. PrestoSpace Deliverable D15.3 MDS3, 2005.

4. Werner Bailer and Peter Schallauer. The Detailed Audiovisual Profile: Enabling Interoperability between MPEG-7 based Systems. In *12th International Multi-Media Modelling Conference (MMM'06)*, pages 217–224, Beijing, China, 2006.

5. Werner Bailer, Peter Schallauer, Alberto Messina, Laurent Boch, Roberto Basili, Marco Cammisa, and Borislav Popov. Integrating audiovisual and semantic metadata for applications in broadcast archiving. In *Workshop Multimedia Semantics - The Role of Metadata (Datenbanksysteme in Business, Technologie und Web, Workshop Proceedings)*, pages 81–100, Aachen, DE, Mar. 2007.

6. Werner Bailer, Harald Stiegler, and Georg Thallinger. Automatic metadata editing using edit decisions. In *Proceedings of 3rd European Conference on Visual Media Production*, London, UK, Nov. 2006.

7. Douglas Bankston. ASC Technology Committee Update. *American Cinematographer*, (12), Dec. 2006.

8. BBC. Standard Media Exchange Framework (SMEF) Data Model 1.5, 2000.

9. Digital Cinema System Spec. v.1.1. Digital Cinema Initiatives, LLC, 2007.

10. DCMI. Information and documentation—The Dublin Core metadata element set. ISO 15836, 2003.

11. Beth Delaney and Brigit Hoomans. Preservation and Digitisation Plans: Overview and Analysis. Technical Report Deliverable 2.1 User Requirements Final Report, PrestoSpace, 2004.

12. DMS-1. Material Exchange Format (MXF) – Descriptive Metadata Scheme-1. SMPTE 380M, 2004.

13. DPX. File Format for Digital Moving-Picture Exchange (DPX), Version 2.0. SMPTE 268M, 2003.

14. Exif. Digital Still Camera Image File Format Standard (Exchangeable image file format for Digital Still Camera: Exif). JEIDA-49, version 2.1, 1998.

15. Roberto Garcia and Oscar Celma. Semantic Integration and Retrieval of Multimedia Metadata. In *5th International Workshop on Knowledge Markup and Semantic Annotation (SemAnnot'05)*, Galway, Ireland, 2005.

16. SMPTE Engineering Guideline. Material Exchange Format (MXF) – MXF Descriptive Metadata. SMPTE EG 42, 2004.

17. Alan Hanjalic. *Content-based analysis of digital video*. Kluwer Academic Publishers, 2004.

18. Jane Hunter. Adding Multimedia to the Semantic Web - Building an MPEG-7 Ontology. In *First International Semantic Web Working Symposium (SWWS'01)*, Stanford, California, USA, 2001.

19. ISAAR(CPF). International Standard Archival Authority Record for Corporate Bodies, Persons, and Families, Second edition, 2004.

20. Chitra L. Madhwacharyula, Marc Davis, Philippe Mulhem, and Mohan S. Kankanhalli. Metadata handling: A video perspective. *ACM Trans. Multimedia Comput. Commun. Appl.*, 2(4):358–388, 2006.

21. MPEG-21. Part 7: Digital Item Adaptation. ISO/IEC 21000, 2004.

22. MPEG-7. Multimedia Content Description Interface. ISO/IEC 15938, 2001.

23. MPEG-7. Multimedia Content Description Interface – Part 4: Audio, Amendment 1. ISO/IEC 15938-4:2002/Amd 1, 2004.

24. MPEG Requirements Group. MPEG-7 Interoperability, Conformance Testing and Profiling, v.2. ISO/IEC JTC1/SC29/WG11 N4039. Singapore, March 2001.

25. MPEG Requirements Group. Study of MPEG-7 Profiles Part 9 Committee Draft. SO/IEC JTC1/SC29/WG11 N6263, Dec. 2003.
26. MXF. Material Exchange Format (MXF) – File Format Specification (Standard). SMPTE 377M, 2004.
27. Frank Nack. Capture and transfer of metadata during video production. In *MHC '05: Proceedings of the ACM workshop on Multimedia for human communication*, pages 17–20, New York, NY, USA, 2005. ACM Press.
28. Frank Nack, Jacco van Ossenbruggen, and Lynda Hardman. That Obscure Object of Desire: Multimedia Metadata on the Web (Part II). *IEEE Multimedia*, 12(1), 2005.
29. Jacco van Ossenbruggen, Frank Nack, and Lynda Hardman. That Obscure Object of Desire: Multimedia Metadata on the Web (Part I). *IEEE Multimedia*, 11(4), 2004.
30. P_Meta. The EBU Metadata Exchange Scheme. EBU Tech 3295, v1.2, 2005.
31. Borislav Popov, Atanas Kiryakov, Angel Kirilov, Dimitar Manov, Damyan Ognyanoff, and Miroslav Goranov. KIM - semantic annotation platform. In *International Semantic Web Conference*, pages 834–849, 2003.
32. Simone Santini and Ramesh Jain. Beyond query by example. In *IEEE Second Workshop on Multimedia Signal Processing*, pages 3–8, Dec. 1998.
33. Data Encoding Protocol using Key-Length-Value. SMPTE 336M, 2001.
34. Metadata Dictionary Registry of Metadata Element Descriptions. SMPTE RP210.8, 2004.
35. Cees G.M. Snoek and Marcel Worring. Multimodal video indexing: A review of the state of the art. *Multimedia Tools and Applications*, 25(1):5–35, Jan. 2005.
36. Raphal Troncy. Integrating Structure and Semantics into Audio-visual Documents. In 2nd *International Semantic Web Conference (ISWC'03)*, pages 566–581, Sanibel Island, Florida, USA, 2003.
37. Raphal Troncy, Werner Bailer, Michael Hausenblas, and Martin Höffernig. VAMP: Semantic Validation for MPEG-7 Profile Descriptions. Technical Report INS-E0705, (CWI) Centrum voor Wiskunde en Informatica, 2007.
38. Raphal Troncy, Werner Bailer, Michael Hausenblas, Philip Hofmair, and Rudolf Schlatte. Enabling multimedia metadata interoperability by defining formal semantics of MPEG-7 profiles. In *Proceedings of 1st International Conference on Semantic and Digital Media Technologies*, Athens, GR, Dec. 2006.
39. Raphal Troncy, Jean Carrive, Steffen Lalande, and Jean-Philippe Poli. A Motivating Scenario for Designing an Extensible Audio-Visual Description Language. In *The International Workshop on Multidisciplinary Image, Video, and Audio Retrieval and Mining (CoRIMedia)*, Sherbrooke, Canada, 2004.
40. Chrisa Tsinaraki, Panagiotis Polydoros, and Stavros Christodoulakis. Interoperability support for Ontology-based Video Retrieval Applications. In 3rd *International Conference on Image and Video Retrieval*, Dublin, Ireland, 2004.
41. Richard Wright and Adrian Williams. Archive Preservation and Exploitation Requirements. Technical Report D2, PRESTO – Preservation Technologies for European Broadcast Archives, 2001.

Part II

Multimedia Semantics

Smart Social Software for Mobile Cross-Media Communities

Ralf Klamma, Yiwei Cao, Marc Spaniol

RWTH Aachen University, Lehrstuhl für Informatik 5, Ahornstr. 55, 52056 Aachen, Germany

The context of a multimedia artifact is neither static nor universally valid. Different people simply interpret and understand contents based on their cultural, intellectual and societal background. While on the Web 1.0 this was already complex to deal with, the Web 2.0 has increased the complexity even more: Web 2.0 communities have become mobile and multimedia-based. The reason is a phenomenon called "Social Software". Social Software simply makes its users to content prosumers (consumer and producer in parallel), anytime and anywhere. Thus, tons of multimedia artifacts are created in one of the many (day by day cumulating) social software applications. However, this is "stupid" as a user can not gain any benefit when trying to (re-)use the so-created contents as Social Software does not "understand" each other. In this chapter we will explore the challenges in making Social Software "smart" by identifying the key issues that are needed to create a software architecture that "speaks" Social Software Esperanto.

1   Introduction

"Stupid is as stupid does'" is a worldly wisdom made famous by Robert Zemeckis' Oscar winning movie Forrest Gump. While the contrary is yet unproven in real life, at least in software the current trend appears to follow the phrase: "Smart is as smart does'". Particularly, the new business models of the Web 2.0 [22] empower media consumers

R. Klamma et al.: *Smart Social Software for Mobile Cross-Media Communities*, Studies in Computational Intelligence (SCI) **101**, 87–106 (2008)
www.springerlink.com

a) by turning them into media and metadata producers (*Data is the Next Intel Inside, Users Add Value, Some Rights Reserved*),
b) by freeing them from the tyranny of the devices (*Software Above the Level of a Single Device*) and
c) by exploiting network effects (*Network Effects by Default, Cooperate, Don't Control*).

Most important professional work is performed in small, trust-based communities of practice (*The Long Tail*). A new generation of professionals called sometimes the Digital Bohemia has been raised up. They organize their professions in highly mobile, situational contexts working on customer sides, organizing meetings in cafés and bars with wireless hotspots, exchange knowledge on events called camps and so on. One of these professionals has compiled a list of social software he is using to organize his professional life in his blog (http://www.zylstra.org/blog), which he is using to tell "what I think about". He uses

> *"Jaiku, what I am doing, Twitter, what I say I am doing, Plazes, where I am and where I was, Dopplr, where I will be, Flickr, what I see, del.icio.us, what I read, Wakoopa, what software I use, Slideshare, what I talk about, Upcoming, where I will attend, Last.fm, what I listen to and then there is my LinkedIn, my Facebook, my Xing, my Hyves, my NING, and my collaborative tools MindMeister, Thinkfold, and Googledocs".*

Concerning these insights there are major challenges to social software that social software is not well prepared for the future on the operational level as well as on the reflective level. Social Software known today as the return of the hacker culture with their mash-ups, API-based interoperability and RSS-based aggregation has to be developed into *community information systems* (or smart social software) which can offer seamless support for complex mobile semantic cross-media processes taking place in the geographically dispersed professional practice of communities.

On the operational level,

- Can we support mobile communities with the collaborative creation of complex multimedia objects?
- Can mobile communities make use of metadata over the frontiers of media, platforms and standards?
- Can we support mobile communities by personalized working and networking strategies in Social Software?

- How do adaptive, mobile web-based interfaces for communities look like?

On the reflective level,

- Can mobile communities continuously elicitate and implement requirements for their professional life? How much computer science support is needed?
- How can mobile communities record their complex media traces and how can they deal with the media traces?
- Can mobile communities maintain or even improve their agency (learning, researching, and working) on the Web 2.0?

In this contribution we concentrate on the operational support of mobile community by introducing our S^3 framework in the following section concentrating on location-based map, multimedia processing and context-reasoning services. In Section 3 we will discuss the architecture for our approach and give some implementation details. In Section 4 we discuss new challenges for mobile communities using such community information systems.

2 The S^3 Challenges and the S^3 Framework

Our *Smart Social Software (S^3)* is a platform-independent lightweight middleware framework, based on a service-oriented architecture (SOA) to provide mobile, web-based community services for heterogeneous communities with diverse but professional requirements. The content and the services can be shared among various tools to support the work of the professionals in communities. The S^3 API is used to build the core functionality of the server. Based on the API the architecture allows open and user-defined server extensions by three basic element types: connectors, components and services. The basic framework has developed comprehensive methods to manage users and groups, permission and roles, and security object management at different levels. Compared to many other approaches, S^3 makes it easier to develop new (smart) social software services for professional communities. The research presented here is guided by three observations about challenges for smart social software.

The *first observation* is that the use of digital media explodes on the Web 2.0 and requirements of community support surge greatly. Social Software is available for any kind of digital media, such as bookmarks, digital images, digital videos etc. The favorite Web 2.0 approach in the moment is a single platform for each digital medium in order to keep system usage as simple as possible. The use of metadata – data about data – is on an incredible level. While a few years ago most researchers still believed that meaningful semantic metadata can only be created automatically, nowadays millions of users create and share so-called *tags* for digital media. Therewith folksonomies emerge for the use of everybody.

While the digital media themselves are in most cases (with the notational exception of digital music) easy to re-use in new contexts, *metadata are still "sticky"*. We want to address two issues here. First, tagging is a practice developed in a community as reification of a shared repertoire. It is not depending on the platform, but on the community. Community support on the aforementioned platforms is only limited to the *single platform itself* and depending on the platform it can be very basic and simplistic. It is almost impossible to re-use semantic metadata in one platform which was originally created in the other platform. Communities cannot use their shared repertoires of reified semantics across different platforms but have to re-create the semantic metadata once again or need special development knowledge to extract the knowledge via APIs and/or RSS.

The fact, that big Web 2.0 companies like Google and Yahoo are buying desperately other Social Software providers buttresses our argumentation because these companies are seeing the same needs. They just follow another strategy ("make it or buy it"). Second however, without any basic agreement on syntactical issues (like using XML) and semantic issues (like using MPEG-7 Semantic Basetypes) sharing of metadata will be almost impossible on a large scale. But sharing of metadata across different platforms is necessary, because *real communities are cross-media communities,* as the example above has demonstrated. They use a mix of different media to reify their current practices and to assign community semantics to digital media especially multimedia with growing bandwidth and computing power. Furthermore, research on how different media settings shape communities and how media changes influence the life cycle and the semantics of communities is still an open issue.

The second observation is that media and space are getting connected again by new technologies like GPS-enabled devices and new emerging practices like geo-tagging. Interaction with smart maps offering advanced location-based map services like geo-tagging, placing, directions, spatial queries and so on are becoming more and more important for professionals in mobile and situational contexts. The utilization of advanced location-based services is still impeded by interoperability issues. There is a heated debate in the Web 2.0 world about lightweight service integration (based on the REST approach) and heavyweight service integration (based on SOAP and XML-RPC). With regard to advanced map services, this discussion is usually leveraged to a discussion of standard-compliant (i.e.

OGC) vs. non-standard-compliant (i.e. Microsoft, Oracle) heavyweight service integration. Usually, it lacks support for cascading web services, and lacks integration of social software application programming interfaces (API) and their web services. It is easy to create a quick mash-up on the API level to have added value by joining e.g. Google maps with a local restaurant guide. It is almost impossible to seamlessly integrate location based web services like route finding and geo-tagging from different providers like Microsoft and Oracle in a complex community scenario on mobile devices.

The third observation is that the environment of social software becomes more and more important. A system is context-aware if it uses context to provide relevant information and/or services to the user, where relevancy depends on the user's task [10]. Social software designers do not know anymore if the user is sitting in front of the stable context of an office-based desktop PC with a LAN-Access to servers. Communities have to consider the use context which consists of the spatio-temporal, the devices and the community context. Besides the common contexts of pervasive systems, community members are switching roles within and between communities while using the same device at the same place. The reason is just that the new Digital Bohemia is working in more dynamic contexts than ever before. On the way to a customer the professionals want to update their media setting to have the newest information from their professional communities, which serves the customers needs best. At the customers' site the professionals want to connect to other community members to get advice or to negotiate ad-hoc meetings. At events the professionals exchange knowledge in a new fashion by making heavy use of social software. In the moment, many activities of this new community interaction are driven more by the enthusiasm of the people than seamless technological support. At the same time, not only professional communities but also travelers, artists, scientists, hobbyists and other communities driven by passion will demand new flexible context-awareness and context-detection services. Our research approach is to provide more support for context-sensing, context-identification, context-matching and context-reasoning by semantic web technologies. Major challenges are the different already existing ontologies for different kinds of contexts. The necessary integration of different context ontologies to apply automatic reasoning techniques is required.

In the following, we discuss smart semantic enhancement, smart mobile community interactions, and smart context handling. We discuss community interactions based on smart maps as we find them nowadays in many mobile social software applications. Semantic enhancements of multimedia content, context awareness and context handling are also crucial for mobility.

2.1 S^3 Semantic Multimedia Enhancement Services

Communities of practice are groups of people who share a concern or a passion for something they do and interact regularly to learn how to do it better [38]. Real communities are cross-media communities that are not restricted to use one single platform for sharing one particular type of digital media. But community members are also part of different communities. Moreover, communities are driven by different needs and have distinguished missions. While in context of community A multimedia contents might be interpreted on a professional level, the context of community B might be leisure time activities. This constraint implies the necessity to switch between different community contexts. Here are two challenges in metadata management: cross-media metadata and cross-community metadata.

Cross-Media Metadata

The difference between technical extraction of data and the semantically correct interpretation of content is called by Santini and Jain, and later Del Bimbo the "semantic gap" [29] [8]. "Simple" metadata annotations only help to give multimedia contents a plain keyword(s) based name. Semantic metadata annotations provide an opportunity to define semantic entities and to assign semantic entity references to a multimedia content. They are more expressive than plain keyword-based descriptions, as they carry additional semantics. Semantic annotations consist of a name, an optional definition, a mandatory type and optional type specific information.

While semantically correct media retrieval is already a challenge in the mono-medial case, the situation is even more challenging in the cross-medial case. One of the reasons is that common search interfaces are mostly restricted to textual queries, regardless of the contents to be searched for. In general, implicit multimedia semantics face the problem of the vocabulary used for tagging [13] [19]. Particularly, the semantics are not accessible for further machine processing. Even more, a holistic approach for cross-media management may not be limited to a single social software application, but requires a comprehensive strategy along the many different social software applications.

The "Multimedia Content Description Interface" ISO/IEC 15938 (MPEG-7) [16] has advanced multimedia description schemes to describe and manage multimedia artifacts as well as multimedia collections. Thus, it offers a comprehensive framework for the management of multimedia contents. However, its overwhelming size leads to problems for users with a limited technical background. Thus, MPEG-7 has not yet been widely used in social software applications. The multimedia semantics of metadata descriptions can not be unfolded as the features of MPEG-7 remain unutilized. In order to unveil the full potential inherent in the MPEG-7 descriptors and description schemes "smart" social software is required. Through these evolving standards multimedia contents are enriched with semantic metadata leading to more advanced multimedia management and retrieval methods to handle the dramatically increasing amount of publicly available multimedia content on the web [2]. Consequently, the annotations themselves should carry their semantics explicitly in order to make this additional information machine-accessible.

In media-centric communities the vocabulary of semantic types is a decisive success factor for S^3, while the MPEG-7 metadata standard is a perfect match. MPEG-7 provides the following seven distinct types for the semantic classification of multimedia contents: *Agent, Object, Place, Time, Event, Concept* and *State.* Each of these seven types allows the specification of additional type-specific information such as geographic coordinates for locations, time points or intervals for time, parameter name/value pairs for states, etc. One prominent problem of plain keyword annotation is the potential risk of semantically ambiguities.

S^3 *Community-Aware Metadata Extensions (Commsonomies)*

The joint semantics of multimedia contents within a community is commonly captured within a so-called "folksonomy" which is created by plain keyword annotations of multimedia contents in Web 2.0 applications. However, it does not reflect the underlying community memberships. In this sense, it is non-community-sensitive. Only if being appropriated in a community context, the semantic enhancement of folksonomies starts to be a true reification of community practices [21]. For example, every user has access to all annotations assigned by all users, possibly within the contexts of different communities. Thus, it is impossible to specify the community context in which an annotation has taken place.

Existing social software applications often neglects that users constitute into groups, which might want to restrict access to group specific media. Even more, depending on their context these communities might use diverse terminologies and might have different viewpoints on the different multimedia artifacts. Thus, different communities should be able to create different community-specific terminologies for multimedia contents. This is the underlying idea of the so-called commsonomies. S^3 however intends to gap this shortcoming by modeling community-aware multimedia annotations using commsonomy. The underlying principle is a "community forests", i.e. a set of hierarchies along with a special notion of community membership semantics. A user thus is given community-specific access to multimedia contents based on his membership to a community. The result can be represented by tree nodes, each of them representing a dedicated (sub-)community. Semantic annotations assigned to a multimedia content can thus be considered in the context of the corresponding community member (e.g. an Administrator, an Expert or a Greenhorn).

Semantic Multimedia Processing

Before how to handle cross-media and cross-community metadata in S^3-CSE is introduced, three typical activities in multimedia processing including community-aware semantics, clustering and retrieval of multimedia contents are discussed. Given these operations we will explain how the previous criteria for S^3 are made productive to unleash the real semantics of multimedia contents in communities.

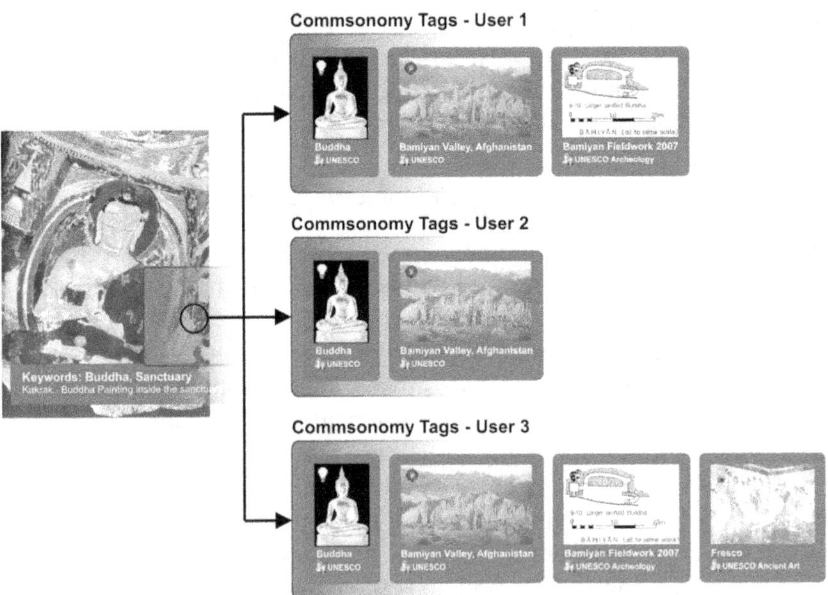

Fig. 1. Commsonomy tag visibility for members of different communities

Community-awareness in multimedia semantics is context dependent. That means semantic annotations are bound to a specific community context (user, community, role, access rights, etc.). Thus, displaying of semantic annotations depends on the context of a particular community membership. Fig. 1 demonstrates community-aware multimedia annotations on client side. Here, three views on the same image based on three different community contexts are shown. In the case that a semantic annotation has been added in a specific community context corresponding to a user's community affiliation, the information is shown. Hence, the semantic annotation is rendered as a thumbnail being part of a multimedia information overlay. Annotations from community contexts that are not related with that particular user are certainly invisible. A desideratum in multimedia retrieval is the missing data quality of solely text or content based multimedia retrieval. Thus, clustering is applied in S^3 in order to reduce the risk of semantic misinterpretation rule-based semantic. Our approach links text and content based multimedia retrieval to achieve a more concise query processing. The k-means clustering algorithm is applied to speed up the query evaluation. By comparing the reference image with the cluster vectors, this procedure has not to be performed with all n images, but only k times with the reference vectors instead (usually $n>>k$)

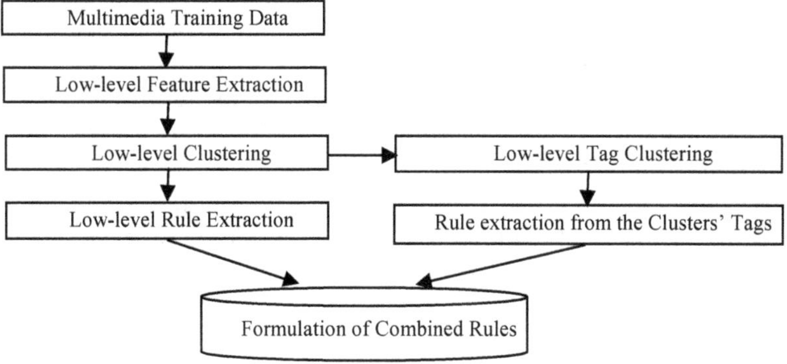

Fig. 2. Rule-based image clustering process

The rule-based clustering process depicted in Fig. 2 works as follows. In an initialization step S^3 the low-level feature vectors of a multimedia test collection are extracted. We only explain the application of multimedia clustering in S^3 for images and videos (based on the extracted key frames). It can be applied on arbitrary multimedia contents in general. At the first stage, a relevant set of multimedia training data are specified for extraction of image feature vectors. Ensuring the interoperability, we utilize those descriptors of the MPEG-7 metadata standard that have proven to be performed on image sets concisely and fast: *Edge Histogram*, *Scalable Color* and *Color Layout*. From these feature vectors we create arbitrary k image clusters, using a modified k-means clustering algorithm. We apply Ward's minimum variance linkage [12] to form preferably homogeneous clusters. The processing results are k clusters represented by the cluster's centroid feature vector. In the next step follow two operations. One is the extraction of low-level rules to express the maximum distance from a centroid's feature vector allowed for images belonging to it. The other is the extraction of tags for each cluster from its members, such as creating a tag-cloud of terms for each cluster. From this tag cloud vector rules are derived for each cluster so that a sub-clustering based on the high-level semantic annotations can be extracted. Finally, both low-level feature rules and high-level tag rules are combined. Thus, the gap between purely low-level content analysis and high-level metadata annotations can be bridged. In order to make the previously extracted rules understandable and interpretable by a reasoner now we describe how the rules are represented in the OWL ontology.

The rules extracted from low-level feature clustering are stored in OWL-classes that contain the specifics about any of the k clusters. The class *LowlevelCluster_k* is the representative of the kth cluster. This class contains the information about the centroid's feature vector as well as the cluster's range *Interval_k$_{min}$_k$_{max}$*. Based on these information it can now be specified whether an image belongs to a certain cluster or not. Similarly, the extracted rules from the clusters' tags can be expressed in OWL-classes. For instance, the class *LowlevelCluster_k_ Tag$_x$* contains the *Tagname* as a value of the xth tag in cluster k (each image and, thus, every cluster may be assigned with more than a single tag). As a result, for each cluster the associated high-level tag are formulated as a rule. In order to apply the inference mechanisms of an OWL reasoner, for each image an instance is being created.

At retrieval, the instances are first queried for a certain *Tagname x*. All clusters which contain this value are then identified. Then, the reference image's feature vector is compared with the cluster's centroid vector. In case the difference is below a pre-defined threshold the dedicated cluster is prepared for the result. Then the result can be delivered to (mobile) clients. More details can be found in [31].

2.2 S³ Smart Map Services for Mobile Community Interactions

With the advent of geo-tagging strategies to connect media and space, map based interaction within mobile communities gains more and more attention. In the moment, Google Maps and Virtual Earth are the basis for a lot of mash-ups to offer added value regarding to the Web 2.0 business models [25]. Professional communities may make use of these mash-ups but usually state more complex interaction patterns such as orchestrating different map-based interaction services within a community process. For example, users often switch views from aerial to bird's eye, to place multimedia objects on maps or to do complex route planning for visiting different sites within one activity. To achieve such complex interaction, on a technical level we need the integration of different map-based services not only in a fixed programming style as mash-ups are offering them right now, but also on the level of semantic composition and orchestration of map-based services. Obvious obstacles today are not only the gap between Web 2.0 API and lightweight web service interaction or heavyweight web service interaction of major geographic information system providers, but also the missing interoperability on the description level between international standards based services and non-standards based. As in the case of multimedia processing services lacking interoperability is mainly the result of lacking use of agreed standards, this is again the core problem here. Standards for geographic infrastructures are applied for location-based service development. A widespread standard is provided by the international industrial GIS consortium, the Open Geospatial Consortium (OGC) [39]. The OGC OpenGIS specification includes four main components: Web Map Server (WMS) Interface Implementation Specification, Web Feature Server (WFS) Implementation Specification, Web Coverage Server (WCS) Implementation Specification and Geography Markup Language (GML). A WMS provides interfaces (API) to simplify the web-based client and server programming for processing map requests. A WFS describe data manipulation operations including insertion, deletion, update, and get and query a geometric object with spatial or non-spatial constraints. A WCS provides access to potentially detailed and rich sets of geospatial information in the form of multi-themes by defining theme manipulation operations.

Besides the standards, there already exist impressive map services on the market. Oracle MapViewer is an OGC compliant web service suite supporting the rendering of data delivered using the OGC WMS protocol. It offers the GetMap, GetCabilities and GetFeatureInfo requests as defined in the OGC WMS Implementation Specification. Microsoft MapPoint Web Services (MWS) is a hosted, programmable SOAP web service suite application developers can use to integrate high-quality maps, driving directions, distance calculations, proximity searches, and other location intelligence into applications, business processes, and Web sites [35]. Microsoft also released the AJAX technology-based Virtual Earth platform to provide mapping features for real-time visualization of location and location based information. Google Maps displays maps with detailed bird's eye and aerial imagery data, using AJAX technologies. Additionally, an open

Google Maps API allows customization of map outputs such as attachment of multimedia onto maps. Users can embed map controls of Google Maps to win a full control over map navigation of streets and imagery data [23].

Requirements for integrating map web services for smart mobile communities are:

- Integration with OGC-compliant map web services
- Integration non OGC-compliant map web services
- Integration with Web 2.0 mash-ups

A comparison among four major heavy-weight web map servers, GRASS (Geographic Resources Analysis Support System) [33], SAGA (System for Automated Geo-scientific Analyses) [28], DEEGREE [9], and CCMS (Cubewerx Cascading Map Server) [1] revealed that none of them is capable of fulfilling all requirements. The conceptual goal of S^3 is to integrate these different highly sophisticated products and to support communities by offering them interactions with smart maps on mobile platforms. Map services are called on the mobile devices through XML which avoids large data transmission.

2.3 S^3 Context-Aware Adaptation Services

For mobile communities not only the spatio-temporal context is important but also the user preferences and the capabilities of the devices used by the people, while the community context in current research is often neglected. Context-aware adaptation meets the requirements of the various users of systems. It is most challenging to create context-aware systems in consideration of more than one single context facet.

Many context-aware systems only consider single contexts and no context modelling approaches are applied at all. Context models are used as internal data structures. But context modelling is a method [3] for common representation of context [15] and to abstract context information in standard formats. The most common context modelling approaches were surveyed for spatio-temporal community context. Based on the survey in [34], we compare key-value based approaches [4], markup scheme approaches [36], graphical model approaches [14], object-oriented model approaches [7], logic-based approaches [18], and ontology based approaches [37] [27].

Generally speaking, ontology-based approaches best fulfil the requirements like distributed composition of contexts, partial (local) validation of contexts, information quality, formality, spatial-temporal and community contexts. Because of a great variety of context information it is difficult to manage a large amount of context knowledge. Thus, ontologies are divided into a common upper ontology and domain-specific ontologies. The upper ontology (cf. Fig. 3) captures general information about the real world in pervasive computing environment such as user context, computing context, physical context, and time.

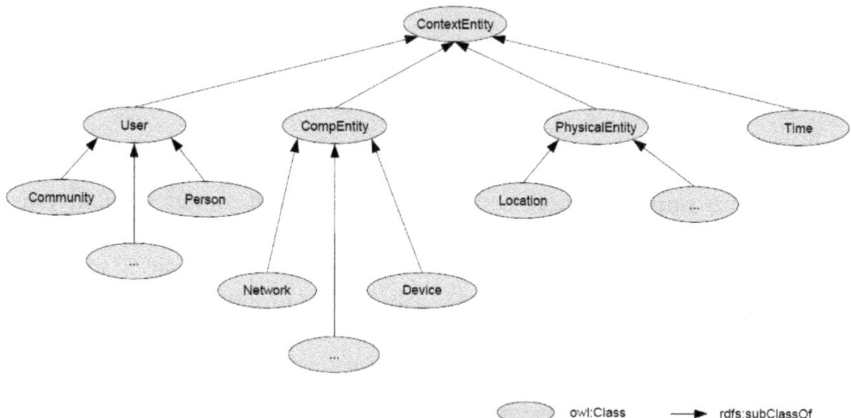

Fig. 3. Upper ontology for context-awareness S^3 extensions

The community ontology models community in terms of various groups. Here, communities are identified by the means of using a clustering algorithm [17]. The instance of the community ontology is created, when a new user comes. There are three key classes in the community ontology. The class `community:Community` contains property `community:Name` and a number of instances of class `per:Person` of the ontology Person. The object property `community:isMemberOf` defines the relationship between `Person` and `Community`. In addition, context information is transferred from the client side to the server side. Context information including GPS coordinates, user's personal information etc. acquired from a client user interface is sent to the *ContextProvider* service on the server side, when a command like "upload context information" is invoked by using the context-aware search. The arriving context information on the server is then represented as instances of ontology `Person`. Together with the ontology `Space`, in which buildings information are stored, the process of location and time reasoning is carried out, finally the results are sent to the user on the client side. SPARQL [32] is the RDF Query language to be used to reason about the context ontology, e.g. about the spatial context (cf. Figure 4).

```
(?p per:spc:IsNearby ?building) <-
(?p per:LocatedAt ?loc1),
(?loc1 geo:longitude ?lon1), (?loc1 geo:latitude ?lat1),
(?building spc:hasCoordinates ?loc2),
(?loc2 geo:longitude ?lon2), (?loc2 geo:latitude ?lat2),
(?loc1 spc:IsNearby ?loc2) .
```

Fig. 4. An instance of SPARQL

3 S^3 Framework Architecture

From the technical point of view the development of smart social software (S^3) therefore implies: smart semantic enhancement, smart context handling and smart mobile community interactions. Fig. 5 shows an excerpt of the S^3 architecture including annotations to illustrate how basic S^3 concepts are used and combined in order to achieve community-aware semantic multimedia annotations.

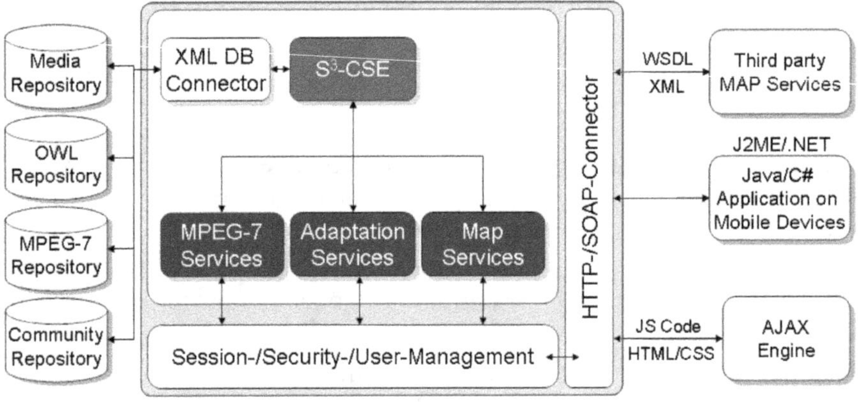

Fig. 5. S^3 architecture with services on the server

We use four different repositories to store XML data for the extended services within the S^3 architecture. The media repository hosted on a FTP server (for easy security ticketing reasons) currently contains the multimedia files themselves. The OWL repository contains the context ontologies for the semantic description of multimedia materials. The metadata about audiovisual data is stored in the MPEG-7 repository. The community repository stores user and community data, roles, access rights and profiles. All repositories are connected to the S^3-services via database connectors, i.e. all services use a built-in S^3-CSE component for the interaction with a native XML database, e.g. eXist [11] [20], IBM DB2 Viper [5], or Oracle

10g [6]. The MPEG-7 Services, the Adaptation Services and the Map Services are specializations of the S^3-CSE service framework [30]. Via the HTTP and the SOAP connectors we have developed gateways to third party Map Services like MS MapPoint and Oracle MapViewer, Java or C# based applications on mobile devices and to an AJAX engine exchanging information with APIs for Google Maps and MS Virtual Earth.

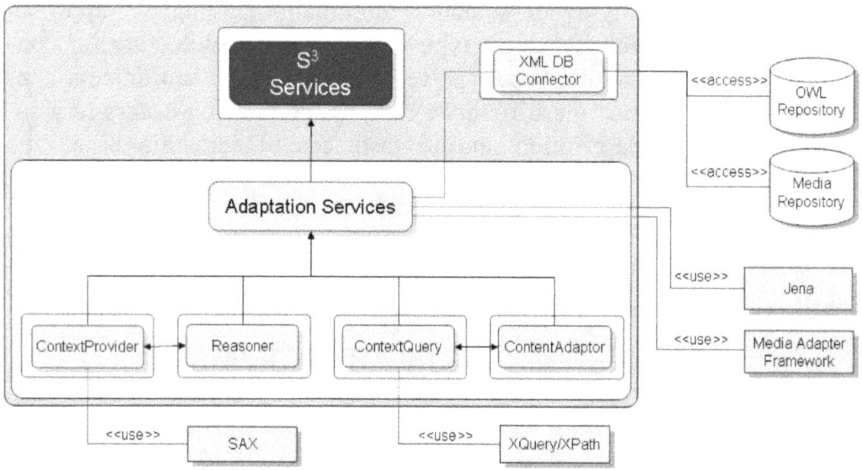

Fig. 6. Detailed design of S^3 Adaptation Services

Three services design and implementation are shortly discussed as follows. Figure 6 illustrates an *adaptation service* extension for a context-aware system. The extension `ContextAwareAdaptationService` is the core service for the context-aware system. It is created for providing functionalities like context acquisition, reasoning, query, and content adaptation. These functionalities are implemented in sub-services such as `ContextProvider`, `Reasoner`, `ContextQuery`, and `ContentAdaptater`. Each sub-service derived from `ContextAwareAdaptaionService` inherits support for access the media repository and the OWL ontology repository via an XML DB-Connector. For representation, reasoning, and query of context ontologies, the Jena framework [24] is used in the `ContextAwareAdaptaionService`, the sub-services therefore inherit those functions in the Jena framework. For developing the content adaptation service, the Media Adaptation Framework from is integrated in the core service. For the implementation of functionalities in the sub-services, some external tools are used, such as SAX, XQuery and XPath for processing XML files in Java.

MPEG-7 Services (cf. Figure 7) in the S^3 architecture are based on two generic multimedia annotation services. The first one is a semantic service for the management of MPEG-7 semantic base type descriptions, while the second is a multimedia content service for the management of MPEG-7-based multimedia content descriptions. Semantic tagging is realized in the multimedia content service by adding semantic base type references to the semantics descriptor of a multimedia content descriptor. In order to create support for community-aware semantic multimedia tagging, we apply an additional custom security object type within the S^3 architecture for controlling access to semantic base type references within a multimedia content description. Notice the difference between controlling access to a semantic base type description and to instances of semantic base type references.

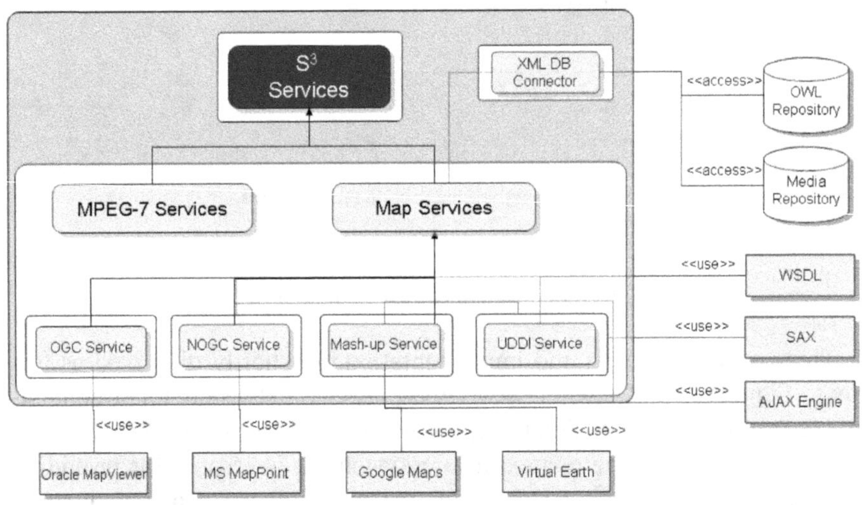

Fig. 7. Detailed design of S^3 MPEG-7 and S^3 Map Services

We offer three different *map services* within the S^3 architecture. First, the OGC-compliant service is represented by e.g. the Oracle MapViewer. The service works on the basis of WSDL descriptions. S^3 Map Services are described according to the WSDL schema and published into a private UDDI registry via the UDDI service. The message transport protocol between clients and services is SOAP over HTTP. Second, the non-OGC Services provides subscription based Web Services that deliver maps, driving directions, distance calculations and other location based services. Because the .NET based services follow the SOAP and WSDL Web Services protocols, we can integrate e.g. MS MapPoint functions into almost any web enabled application, using Eclipse and Apache AXIS. The Mash-up

Service realizes an AJAX engine which is responsible for both rendering the user interface and communicating with the server. When an application invokes the Mash-up Service in the S^3 Map Service, it sends JavaScript strings back to the application as results. Then, the application calls the AJAX Engine using those JavaScript codes. With the AJAX engine, JavaScript will communicate directly with Virtual Earth or Google Maps servers, using the JavaScript XMLHttpRequest object. After getting the response, AJAX engine sends back HTML or CSS data to the application and the application will render those data as new map images on the screen.

4 From Smart Social Software to Smart Communities

The implementation of the services for the S^3 architecture is a technical adventure now. Although most software developers concerned were quite satisfied with using the S^3 framework, there were still many problems to be solved. Database issues were mainly overcome by implementing a generic XML DB component to handle different XML databases uniformly. XML database design, querying and updating are still a nightmare for developers. Although we use standard JDBC or ODBC technologies the subtle differences between different databases are still disputable. On the middleware level there are a zillion of available tools. Integration of WSDL generators, UDDI registries, reasoner frameworks, OWL tools and so on was a major task for all developers. Even worse is the situation on the client side. We tried also to provide user interfaces for these applications but limited processing capabilities on mobile devices, e.g. in XML processing challenged the software developers in addition.

In the future we will even integrate tighter multimedia processing and semantic processing in our community-specific way. Those services will be demonstrated on mobile devices. New standards on the horizon like ramm.x [26] show that we are heading in the right direction. Some services have been realized in the last year.

Having worked on such a basis, we are now exploring how to compose and orchestrate the services not only for offering smart services for mobile communities but for having smarter communities by integrating non-human agents in the community practices or by combining services. Communities of practice are not self-contained entities, but develop in large historical, social, cultural, and institutional contexts. The contexts result in specific resources and constraints [38]. In S^3 framework, we defined the context services in terms of spatial, temporal and device contexts. The context services can provide communities adaptability and flexibility and even more mobility. For example, how to integrate context reasoning over ontologies and MPEG-7 metadata descriptions?

Social software provides communities a platform to create, search and share media by different means. One of the effective methods, tagging is seen as an opposite approach against the Semantic Web and ontology. However, smart social software enables communities to do tagging in different ways including community aware tagging, MPEG-7 semantic tagging and geo-tagging, which builds up ontology. Thereafter, context reasoning can be performed over ontologies. While ontology-based reasoning technologies do not scale in Social Software contexts for millions of potential users they do fit in community-centered processing of semantic information. There are a lot of available solutions we can make use right now making communities smarter.

In conclusion, S^3 framework hosts mobile communities with smart social software. Based on observations of future use in mobile communities of professionals we have designed services for enhancing multimedia processing by integrating Web 2.0 features (commsonomy). We have combined metadata strategies for improving multimedia retrieval, map based services and context aware multimedia handling. Most of the services explained here are just experimental and implemented for the first time ever in a community engine. More important are conceptual and architectural considerations which have demonstrated the S^3-CSE to be an extensible and scalable environment for realizing new multimedia enhanced services for communities.

References

1. J. Beaujardiere. OpenGIS Consortium Web Mapping Server Implementation Specification 1.3. *OGC Document no. 04-024*, August 2002.
2. A. Benitez, H. Rising, C. Jörgensen, R. Leonardi, A. Bugatti, K. Hasida, R. Mehrotra, A. Tekalp, A. Ekin, T. Walker. Semantics of Multimedia in MPEG-7, in: *IEEE International Conference on Image Processing, Vol. 1, pp. 137-140*, 2002.
3. S. Brinkkemper, K. Lyytinen, and R. Welke (eds.). Principles of Method Construction and Tool Support. *Springer-Verlag*, 1996.
4. G. Chen, and D. Kotz. A Survey of Context-Aware Mobile Computing Research. *Technical report, Dartmouth College, Hanover, NH, USA,* 2000.
5. W.-J. Chen, A. Sammartino, D. Goutev, F. Hendricks, I. Komi, M.-P. Wie, R. Ahuja. DB2 9 pure-XML Guide. *IBM Redbooks*, ISBN 0738489832, 2007.
6. M. Cyran. Oracle Database Concepts, 10g Release 2 (10.2). *B14220-02, Oracle*, October, 2005.

7. K. Cheverst, N. Davies, K. Mitchell, and A. Friday. Experiences of developing and deploying a context-aware tourist guide: the GUIDE project. In: *MobiCom'00: Proceedings of the 6th annual international conference on Mobile computing and networking, 2000. ACM Press*, pp 20–31, 2000.

8. A. Del Bimbo. Visual Information Retrieval. *Morgan Kaufmann*, 1999.

9. DEEGREE Team. DEEGREE Project. September 2006, http://deegree.sourceforge.net/ (last access: October 19, 2007)

10. A. K. Dey. Providing Architectural Support for Building Context-Aware Applications. *PhD thesis, Georgia Institute of Technology*, 2000.

11. eXist. An Open Source Native XML Database. www.exist-db.org, 2007 (last access: June 1, .2007).

12. D. Fasulo. An Analysis of Recent Work on Clustering Algorithms. *Technical Report 01-03-02, Department of Computer Science and Engineering, University of Washington, Seattle, WA 98195*, April 1999.

13. G. W. Furnas, T. K. Landauer, L. M. Gomez, and S. T. Dumais. The vocabulary problem in human-system communication. *Communications of the ACM, 30(11):964-971*, 1987.

14. T. Halpin. Information Modeling and Relational Databases: From Conceptual Analysis to Logical Design. *Morgan Kaufman Publishers, SF, USA*, 2001.

15. A. Held, S. Buchholz, and A. Schill. Modeling of Context Information for Pervasive Computing Applications. In: *6th World Multiconference on Systemics, Cybernetics and Informatics (SCI2002), Orlando, FL, USA*, 2002.

16. ISO/IEC: Information technology – Multimedia content description interface – Part 5: Multimedia description schemes. *15938-5:2003, ISO*, 2003.

17. R. Klamma, M. Spaniol, and Y. Cao. Community Aware Content Adaptation for Mobile Technology Enhanced Learning. In: *Proceedings of the 1st European Conference on Technology Enhanced Learning (EC-TEL 2006), Hersonissou, Greece, October 1-3, LNCS 4227, Springer, pp. 227-241*, 2006.

18. J. McCarthy and S. Buvac. Formalizing Context (Expanded Notes). In: Buvac, S. and Iwanska L. (Eds.) *Working Papers of the AAAI Fall Symposium on Context in Knowledge Representation and Natural Language, Menlo Park, California,. AAAI, pp. 99–135*, 1997.

19. C. Marlow, M. Naaman, d. boyd, D., and M. Davis. HT06, tagging paper, taxonomy, flickr. academic article, to read. In: *Proceedings of the seventeenth conference on Hypertext and hypermedia, pp. 31-40*, 2006.

20. W. Meier. eXist: An Open Source Native XML Database. In: *Web, Web-Services, and Database Systems, NODe 2002 Web and Database-Related Workshops, Erfurt, Germany, October 7-10, Revised Papers, volume 2593 of LNCS, Springer-Verlag, Berlin Heidelberg, pp. 169 – 183*, 2002.

21. P. Mika. Ontologies Are Us: A Unified Model of Social Networks and Semantics. In: *Proceedings of the ISWC 2005, pp. 522-536*, 2005.

22. T. O'Reilly. What Is Web 2.0 - Design Patterns and Business Models for the Next Generation of Software. http://www.oreillynet.com/pub/a/oreilly/tim/news/2005/09/30/what-is-web-20.html (last access: October 18, 2007).

23. E. Pimpler. Introduction to Developing with Google Maps. March 2006, http://www.directionsmag.com/printer.php?article_id=2120 (last access: October 19, 2007)

24. S. Powers. Practical RDF. *O'Reilly, July* 2003.

25. ProWeb. www.programmableweb.com. 2007(last access: August 31, 2007).

26. Ramm.x. RDFa-deployed Multimedia Metadata (ramm.x) Specification. http://sw.joanneum.at/rammx/spec/ , 2007 (last access: October 5, 2007).

27. I. Roussaki, M. Strimpakou, N. Kalatzis, M. Anagnostou, and C. Pils. Hybrid Context Modeling: A Location-based Scheme Using Ontologies. In *PerCom Workshops, pp. 2–7*, 2006.

28. SAGA Team. SAGA Project. September 2006, http://www.saga-gis.uni-goettingen.de/html/index.php?newlang=eng (last access: October 19, 2007)

29. S. Santini and R. Jain. Beyond Query by Example, *ACM Multimedia 1998, pp. 345-350*, 1998.

30. M. Spaniol, R. Klamma, H. Janßen, and D. Renzel. LAS: A Lightweight Application Server for MPEG-7 Services in Community Engines. In: *Proceedings of I-KNOW '06, 6th International Conference on Knowledge Management, Graz, Austria, September 6 - 8, 2006, Springer, pp. 592-599*, 2006.

31. M. Spaniol, R. Klamma, M. Lux. Imagesemantics: User-Generated Metadata, Content Based Retrieval & Beyond. *K. Tochtermann, H. Maurer, F. Kappe, A. Scharl (Eds.): Proceedings of I-Media'07, International Conference on New Media Technology and Semantic Systems, Graz, Austria, September 5 - 7, 2007, J.UCS (Journal of Universal Computer Science) Proceedings, pp. 41-48*, 2007.

32. SPARQL Query Language for RDF. *W3C Candidate Recommendation*, 14 June 2007, http://www.w3.org/TR/rdf-sparql-query/ (last access: October 19, 2007)

33. A. Sorokine. Implementation of a Parallel High-performance Visualization Technique in GRASS GIS. *Computational Geoscience*, 33 (5):685-695, May 2007.

34. T. Strang and C. Linnhoff-Popien. A Context Modeling Survey. In *First International Workshop on Advanced Context Modelling, Reasoning and Management at UbiComp*, Nottingham, UK, Sept. 2004.

35. C. Thota. Programming Mappoint in .NET, *O'Reilly Media, Inc, ISBN:0596009062*, 2005.

36. W3C. Composite Capabilities / Preferences Profile (CC/PP)., http://www.w3.org/Mobile/CCPP, 2006 (last access: October 19, 2007)

37. X. H. Wang, D. Q. Zhang, T. Gu, and H. H. Pung. Ontology Based Context Modeling and Reasoning using OWL. In *PERCOMW '04: Proceedings of the Second IEEE Annual Conference on Pervasive Computing and Communications Workshops, IEEE Computer Society, pp. 18 ff*, 2004.

38. E. Wenger. Communities of Practice: Learning, Meaning, and Identity. *Cambridge University Press, Cambridge, UK*, 1998.

39. A. Whiteside (ed.) OpenGIS web services architecture description, 0.1.0, Nov. 2005, http://www.opengeospatial.org/standards/bp (last access: October 19, 2007)

Organizing metadata into models and ontologies for lowering annotation costs of a biological image database

Arnaud Da Costa[1], Eric Leclerq[1], Arnaud Gaudin[2], Jean Gascuel[2], and Marie-Noelle Terrasse[1]

[1] LE2I UMR CNRS 5158 - `firstname.lastname@u-bourgogne.fr`
[2] CESG UMR CNRS-UB-INRA 5170 - `lastname@cesg.cnrs.fr`

Summary. In this chapter we present the initial phase towards lowering annotation costs in a specialized image database application. The proposed coupling of model-based description with a domain-specific ontology allows to use metadata during the annotation process. This makes creating annotations feasible even for domain technicians while guaranteeing consistency of annotations from an expert's point of view.

1 Introduction

Information Systems (ISs) are often built from scratch [31, 34]. This results into a large number of ISs, each of them using a different way to handle close concepts of knowledge representation. With the introduction of UML 2.0 and MOF 2.0, the OMG[3] proposed to organize domain descriptions in terms of MOF-repositories which contain metamodels and related models. This model-based approach leads to metamodeling architectures in which metamodels describe application domains and models describe applications. Yet, model-based descriptions are fragile in the sense that the categorization they induce (i.e., the grouping of objects into categories) can be questioned. A model can satisfy all the specified requirements while its resulting classification of objects does not satisfy all of the constraints that are "hidden" within the domain metadata. Such a risk is not acceptable for metadata-intensive applications that require efficient management of complex data as well as their complex metadata. Yet, model management has been recognized as a sound approach to the development of such metadata-intensive applications [3, 18]. Ontologies provide a well-defined mechanism for representing domain knowledge by referring to a set of concepts or terms that can be used to describe some area

---

[3] OMG: Object Management Group `http://www.omg.org`.

A.D. Costa et al.: *Organizing metadata into models and ontologies for lowering annotation costs of a biological image database*, Studies in Computational Intelligence (SCI) **101**, 107–126 (2008)
www.springerlink.com

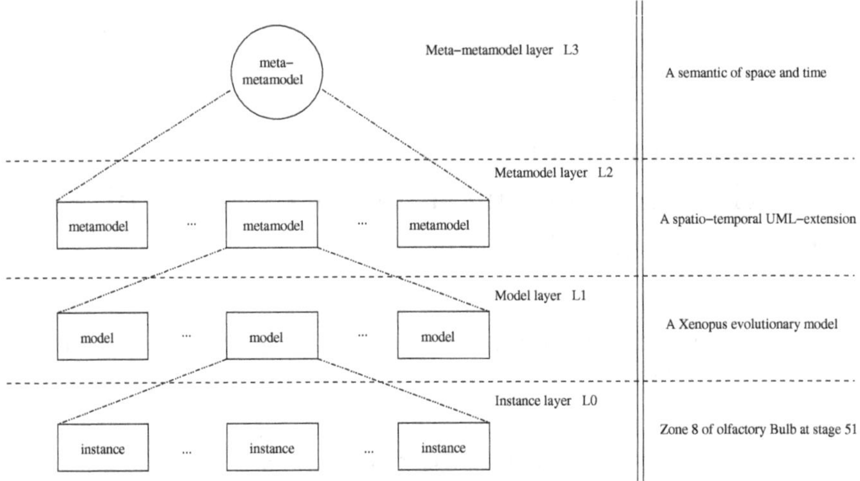

Fig. 1. UML-based four-layer metamodeling architectures [30]

of knowledge or to build a representation of it [27]. Building an ontology is subjected to a choice along two axes [25]:

- horizontal (relevance) axis corresponds to determining the extent of information that should be included in the domain knowledge representation. When representing the *Xenopus* domain knowledge, relevance is about deciding whether to include or not to include a representation of what Xenopus eats (i.e., anything in the aquatic environment: invertebrates, crustaceans, amphibians, and fish).
- vertical (granularity) axis corresponds to determining how fine-grained the representation should be. In the *Xenopus* domain knowledge, granularity is about deciding whether to refine or not to refine knowledge down to molecular level.

Furthermore, ontologies are subjected to a major limitation in the sense that automatic processing of domain knowledge makes sense only if the domain knowledge is consensual [10, 19].

Due to the above restrictions, we assume that most ontologies lie at levels L1 (Model Layer), and L2 (Metamodel layer) of *UML-based four-layer metamodeling architectures* (Figure 1).

Ontologies allow formulation of rules pertaining to their own concepts and they also provide tools for reasoning. Such tools allow to verify that the represented domain knowledge is consistent and they also allow to perform computations and to verify selected elements of the ontology (this is a valuable help if the domain knowledge is to evolve fast: automatic checks can be performed).

Coupling of metamodels and ontologies has been discussed by various authors whose approaches can be classified into two main groups. First, metamodeling architecture can be used to define ontology modelers or to modify definitions of ontology languages (e.g., UML and OWL-DL [6], UML and RDFS [17, 22]). Second, UML itself can be used for ontology development [1, 7, 8] (a detailed state of the art can be found in [14]). Since a model is intended only for a specific application, it doesn't have to be consensual: a model can use constraints that are more specific than those used by ontologies. A model is bound to specific layers (vertical granularity), but unlike ontologies it can cover multiples areas (horizontal relevance): a given application uses information from its main domain and may complement it with the knowledge from some other domains (e.g., a Xenopus-related application might need the domain knowledge of its aquatic environment).

Both ontologies and modeling architectures are useful for interoperability and reuse. The represented domain knowledge can be accessed by different ISs, each IS mapping the applicable concepts into its model. Similarly, reuse of knowledge descriptions can be improved both at the level of semantics (e.g., using formal ontologies such as BFO [24, 25]) and at the level of exchange formats (e.g., a format whose specification is open, such as OWL[16] proposed by W3C under the *W3C Document Notice/License*[33]). The use of ontologies brings out one interesting point for browsing and searching: the ontologic hierarchy can provide a worthy help with querying the Xenopus application: if a query doesn't return any result, it can be useful to generalize one or multiple of the query's concepts, or to use a sibling concept (under certain conditions) to find nearest partial matches[19].

In our proposal, ontologies play three main roles. First, they facilitate categorization of an application model by defining domain attributes as parts of the related ontology. Second, ontologies guide the annotation process by defining patterns and rules. Third, ontologies improve annotation consistency by defining rules that control the annotation process (so that experts can play their role of decision makers).

This chapter is organized as follows. Section 2 outlines our ongoing work on an image database: the *Xenopus* Database. The *Xenopus* collection of images (created at the "Centre Européen des Sciences du Goût") needs to be organized into an image database, with efficient retrieval mechanisms based on both metadata and image content. We describe our methodology and our current results. Section 3 presents our upcoming work in what exactly is being extended.

2 An example application: the *Xenopus* database

The main goal of our biology partners is to understand reorganization of *olfactory neural networks* during the postnatal life. The biology partners use the *Xenopus laevis* biological model, an amphibian extensively used in developmental biology over the last 20 years. Interestingly, this animal has two *olfactory systems*, one involved in olfactory perception in water and the other

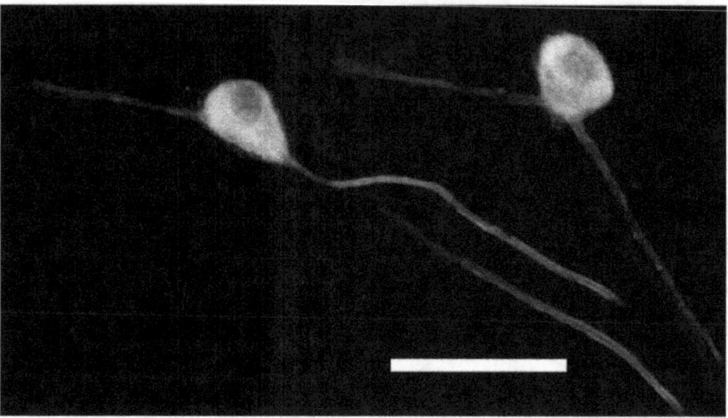

Fig. 2. This image illustrates the morphology of 2 *bipolar neurons* in the *trigeminal olfactory system*. The *neurons* have been *stained* with *carbocyanin DiI*. *Cell bodies* of the two *neurons* and their *axons* and *dendrites* are distinguishable ([11])

involved in olfactory perception in air. The aerial olfactory system appears during metamorphosis and induces a deep reorganization of the *tadpole olfactory system* which is only composed in the *tadpole* state of the *aquatic olfactory system*. In order to analyze how these two different olfactory systems evolve during this reorganization, the partners extensively use imaging technologies. The peripheral olfactory system (nose) and the central olfactory system (brain) have been studied by classical neuro-anatomical techniques involving *dissection, sectioning, staining* and acquisition of images of the corresponding organs.

Different staining protocols have been used, such as immunological tools (*antibodies*), *lectins, carbocyanines*, with each being specific to a particular component of the system. The target of these different markers could be either *glial cells* or neuronal components such as *axons, dendrites, nerves* and *tracts*.

The result of this process of experimentation is a tremendous amount of images with a peculiar diversity which comes from:

1. Different parts of the system observed (Figures 2 and 5 show different organs marked by using the identical staining),
2. The different types of cells,
3. The different *staining* protocols (Figures 5 and 6 show the identical zone of the *Xenopus olfactory bulb* at *stage 59*, with *staining* exhibiting or hiding different elements: *zone 9* is only visible using the *DiI staining*),
4. The different acquisition techniques used,
5. Evolution stages of *Xenopus* (Figures 6 and 7 show the identical *Xenopus olfactory bulb* zone in different *stages*),
6. And the lighting mode used (e.g., *Xenopus* can be illuminated from above by direct lighting or from below by epifluorescence)

In this section we describe the context of our work (Section 2.1), then the *Xenopus* collection of images and extraction of information from these images (Section 2.2), and finally the *Xenopus* ontology (Sections 2.3 and 2.4).

2.1 Context of our work

In order to design and to implement an efficient image database for the *Xenopus* application, we decided to extend a model that had been proposed by Besson [4] for his IkoSem Content Based Image Retrieval framework (a framework for design and testing of image databases created from a collection of images). The Ikosem framework is centered around a generic image description model that is to be instantiated when we wish to build an image database.

IkoSem is based on both physical and semantical features of images: physical features are automatically derived from a set of pixels of an image while semantical attributes are deduced from relationships between an image and the domain terminology. Ikosem's model allows to decompose an image into zones (which are physical parts of an image) and objects (which are image parts that refer to elements of the domain terminology). Three types of relations between image parts are defined: composition (of image parts to build the image in its entirety), spatial (which refers to the 2D space associated with the image) and semantical (which refers to the domain terminology).

Within IkoSem's model, an image is described in terms of *simple objects* (visible in an image or automatically extractable from an image) and of *complex objects* (i.e., combinations of simple objects). Among objects of an image, we can define several relations: composition relations that depict an image as a hierarchy of objects, spatial relations (distance, direction, topological relations), and semantical relations (i.e., relations that refer to the domain knowledge). It is possible to have several composition relations for a given image: each of them corresponding to a different user perspective on the image.

We will define a *pattern* as a list of objects, their respective relations and optional precision for these relations; e.g., a *pattern* can be a list of objects and their relative positions (specified within a narrow margin). Patterns can be attached to images. Patterns contain objects related by spatial relations (i.e., topological relations). Metadata items are attached to images, patterns, and their components.

A domain metamodel was defined for IkoSem (see Figure 3). Such a metamodel defines image parts (an image itself, objects, and zones), their possible attributes (syntactical and semantical attributes), and possible relations among image parts (semantical, spatial, and composition relations).

A model of the Xenopus image database is proposed in Figure 4 which refers to the IkoSem metamodel. In this model, image descriptions contain image parts (i.e., objects) and patterns contain components.

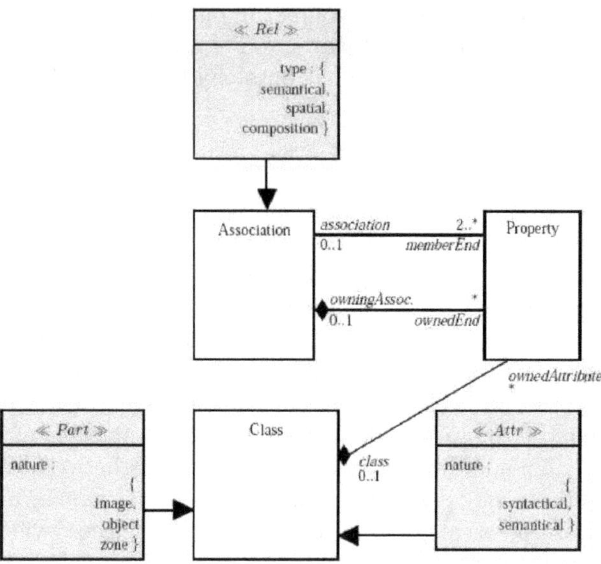

Fig. 3. Domain metamodel defined for Ikosem

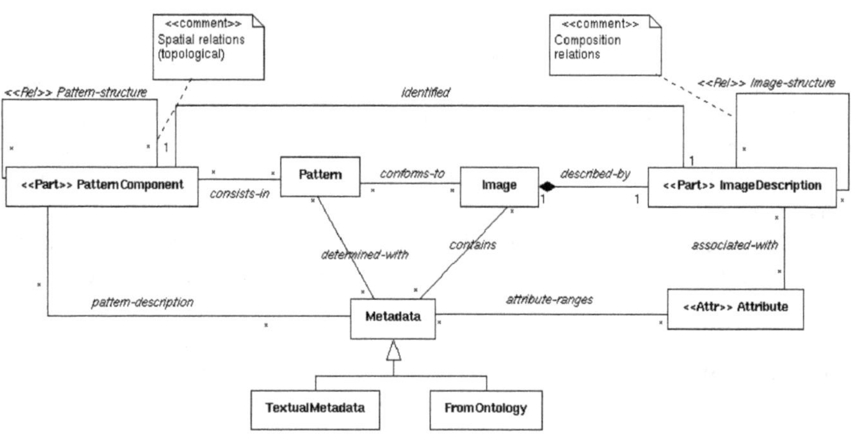

Fig. 4. Model of the **Xenopus** image database

2.2 Extracting information from the *Xenopus* collection of images

Since images in the database can be very different from each other (e.g., Figures 2, 5, 6, and 7), after a consultation with the biologists we decided to focus on the wider homogeneous sub-collection images that are similar to the one of Figure 8). Our first objective was to define and evaluate a set of tools suitable for extracting image parts from the set of pixels of the image and producing annotations. As expected, we had the following two options:

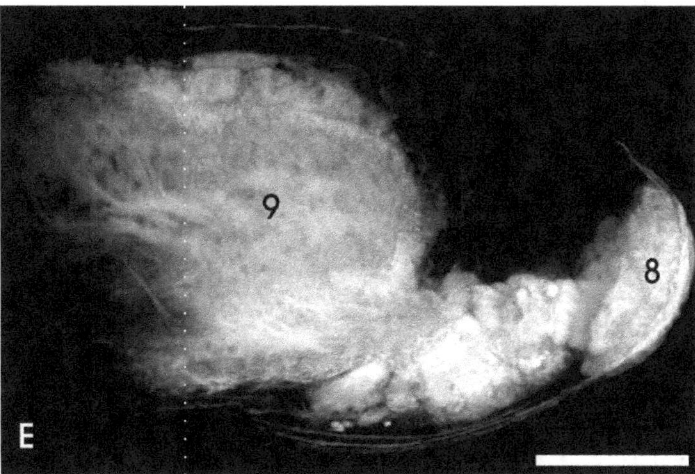

Fig. 5. In this frontal view, the *staining (DiI)* identical to that of Figure 2 has been used but the area of the brain is different. This picture corresponds to the different *zones* of the *olfactory bulb* at the developmental *stage 59* ([11]): *zone 8* is the accessory olfactory bulb; *zone 9* is the air olfaction dedicated zone at this *stage*

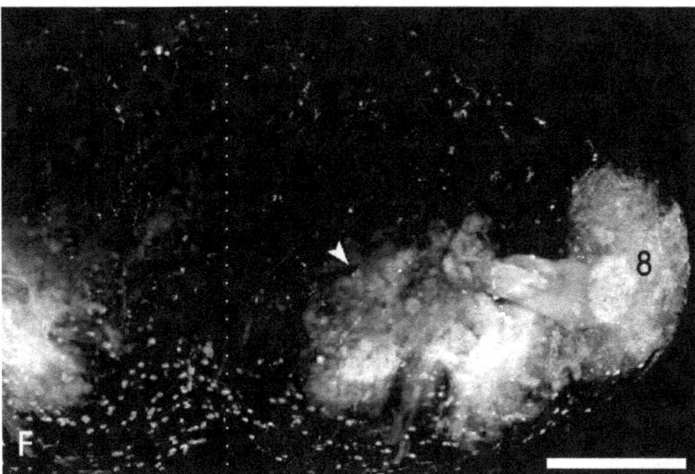

Fig. 6. This picture shows the brain area identical to those of Figure 5, but a different *staining* is used (*lectin*). The developmental *stage* is 59; *zone 8* is distinguishable: the accessory olfactory bulb ([11])

- the first option corresponds to a major effort during the image segmentation phase. This could give us some chance to be able to determine zones by automatic means. Segmentation experts were rather definite on the timing issue: it might take a very long time to define a specific extractor for such a subset of images in the image collection.

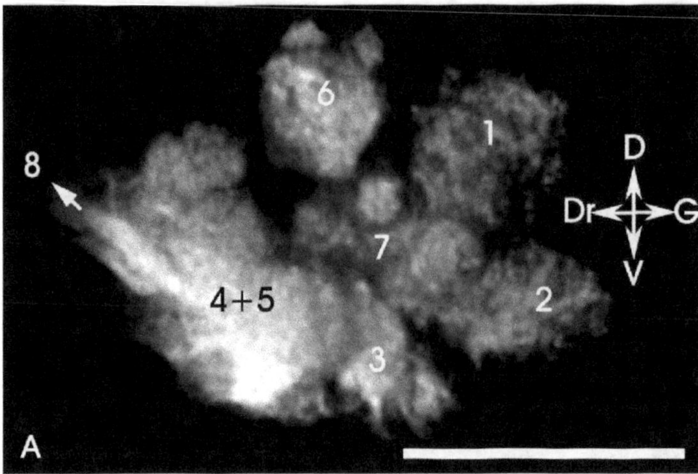

Fig. 7. In this picture the *brain area* and the *staining method* are identical to those of Figures 2 and 5, but the developmental *stage* differs (*stage 44* vs *stage 59*) [11]. At this stage both *zones* are dedicated to aquatic olfaction; zone 1 to 8 exist even though *zone 8* is outside the scope of this picture

- the second option corresponds to making only a limited investment into the image segmentation phase by using a Snake-like algorithm in order to detect zone borders. Such algorithms (e.g., [15]) are expected to segment each image within reasonable processing time and could also be adapted for our images in an acceptable time (using a semi-automatic initialization).

In both cases, we expected a huge amount of post-segmentation work needed for image annotation. Yet there was no reason to believe that accurate segmentation of images would facilitate such annotation. We decided to pursue an alternative, namely to abandon the objective of a fully automated extraction and instead to make writing annotations easy enough even for non-expert technicians. We thus determined three distinct roles a user can play:

Domain technicians are people that annotate images with information that can then be automatically processed. The domain technicians typically have only limited knowledge of the domain (e.g., they may have been trained to recognize some patterns in a picture without having particular knowledge of the underlying concepts).

Domain experts are people that master the domain knowledge. If something is ambiguous, they are the only ones that can make a decision. Domain experts are also the ones that can make new discovery that will impact on the domain knowledge; whether by adding new information to the domain knowledge or by declaring obsolete some concepts of the domain knowledge.

Specialists of the knowledge representation serve as mandatory intermediaries between the domain experts and the computer approaches to representation of knowledge (an ontology in our case). These specialists have significant experience with the recognized best practices in building a knowledge representation (e.g., the initial version of the ontology), and with modifications of the knowledge representation (e.g., adding or deleting information, moving a branch in the ontology from a node to another, etc.).

The next section outlines how we used all the pieces of the available metadata to guide and control annotations. For each image we know the development stage of the *Xenopus* involved. Based on the stage information, we can propose a pattern that specifies both the image parts to be localized (zone markers) and their relative locations. Figure 8 presents such a pattern. Patterns can be combined with physical extractors (e.g., a Snake-like algorithm with manual initialization of each zone marker's position).

In order to implement the pattern-based approach, we decided to emphasize the use of metadata, e.g., by defining allowed and forbidden variations of a pattern. As stated in [35]: "Annotations ... should be based on a rich metadata structure in connection with an ontology." This is based on an ontology that is presented in the following section.

2.3 The ontology structure

Since the people that will annotate the *Xenopus* images are French speaking, we used French terms during the prototyping phase of the work (English terms appear as suffixes to each term).

Following the work described in [23], we decided to separate annotation-related concepts from domain-specific ones. We first focused on all of those issues that could be sources of heterogeneity in the image database, and second on all the complementary domain concepts. This results in the following branches in the ontology (see Figure 10):

- Possibles techniques used to acquire an image are located in the branch **Acquisition_Techniques**;
- *Staining* techniques are defined in a branch called **Staining techniques**. Each staining family is localized in a dedicated branch, sharing close properties to exhibit biological elements;
- Processes that could be applied to images are located in a branch called **Image Processing**. For example, every picture in a stack can be useless by itself, yet the combination of all these images can be of great interest;
- The **Cells** branch was placed at the root of the domain knowledge branch to avoid duplicating it many times within the **Xenopus** branch: most elements of the **Xenopus** branch will also belong to the **Cells** class.
- All known *Xenopus*'s biological elements and Xenopus's structure are in the **Xenopus** branch. We were confronted with the problem of elements that do not persist throughout the whole life of *Xenopus* (e.g., barbels);

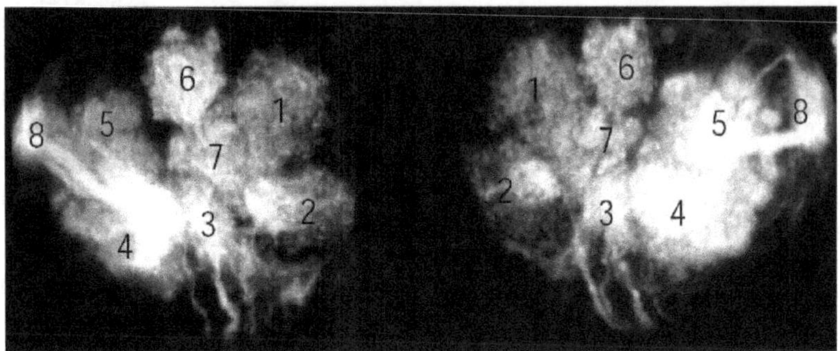

Fig. 8. Frontal view of *axonal* projection in the *olfactory bulb* of *tadpole* at *stage 44* (*DiI staining*, [11])

Metadata	Meaning
Stage is 47	8 *zones* (1 to 8) exist
DiI staining	All *neuron cells* are visible
View is frontal & *Stage* is 47	*zones* 1 to 8 are visible and their relative positions are known

Table 1. Some metadata associated with Figure 8. The information contained in these metadata can provide worthy help with annotations.

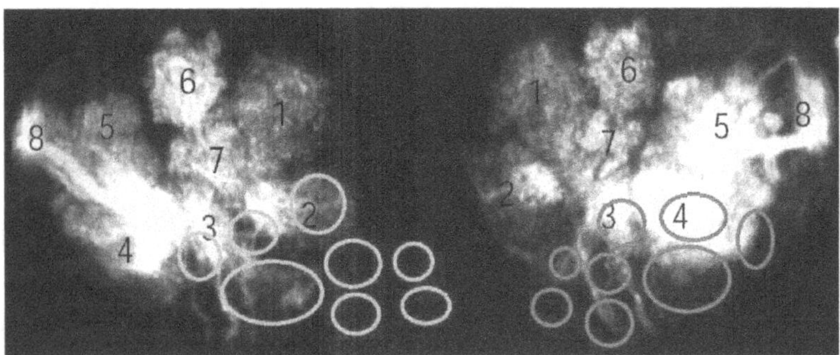

Fig. 9. The ellipses indicate the proposed pattern for a frontal view of axonal projection in the olfactory bulb of a tadpole at stage 47 (DiI staining, [11]). A domain technician will be able to move these ellipses within a margin that depends on the domain rules

Gene Ontology ([21]) chose to distinguish such elements by adding the keyword "sensu", followed by a meaningful text to indicate in which cases they may be used. For example, *barbels* would become *barbels sensu tadpole* in our ontology. We chose to avoid duplication of branches that have similar content by sharing a single sub-tree for all anatomical elements. Properties of the corresponding ontology nodes will be used to determine whether a class may or may not be used for annotation.

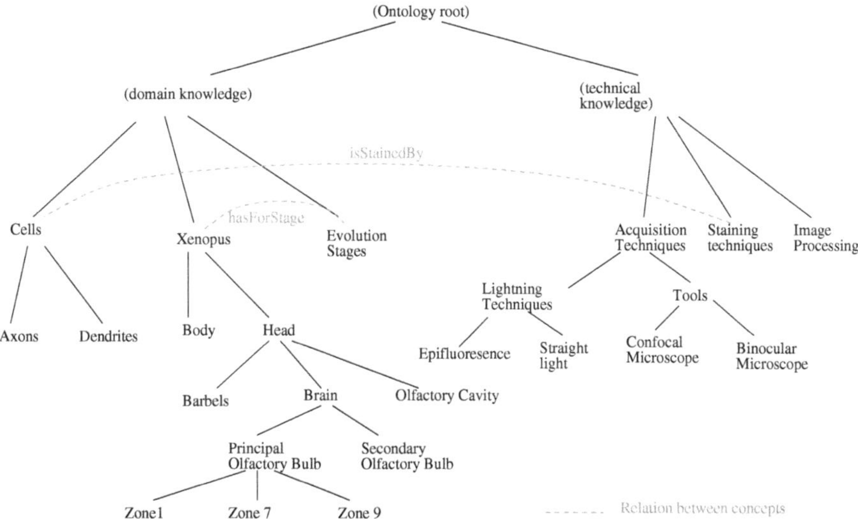

Fig. 10. Hierarchy between some concepts of the ontology, and some relations we set up between different concepts

The ontology was built using the Protg ontology editor and knowledge-base framework [13, 29]. Figure 10 shows some concepts of the ontology and two relations between different concepts, and Figure 11 show partial screenshots of our ontology in Protg.

2.4 The ontology rules and relations

The first part of our work on the ontology building process was centered around extraction-related information and led to the ontology structure described in the previous section. Description of the domain knowledge in distinct branches can be completed with rules and relations. Rules express information supplied by the domain experts; some sample rules are shown in Table 2. An interesting aspect of rules is the possibility to use them as relations between different branches of the ontology. For example, the first rule of

Table 1 states that if *"Stage* is 47", then *"zones* 1 to 8 exist"; this allows to introduce a relation between the **Stage_d_evolution** branch and the **Xenopus** branch[4]. Some of the other families of relations deserve a peculiar interest:

- a **compatibility** relation between **Acquisition_Techniques** and **Staining_Techniques** indicates whether a particular *Acquisition* technique is relevant for a particular *staining*;
- a **visibility** relation between *Cells* and *staining* indicates which cells will be exhibited by a specific staining.

Table 2 lists some of the rules we established with the *Xenopus* domain experts. A relation between *Cells* and **Image_Processing** will soon be extended since some *zones* are visible only if an appropriate image processing has been performed (e.g., a stack combination to reveal *zones*).

Xenopus in stages between 1 and 28 is an embryo
Xenopus in stages between 28 and 66 is a tadpole
Xenopus in stages higher than 66 is an adult
a *Xenopus* in stage x where $1 < x < 66$, might also be annotated with stage $x - 1$ or stage $x + 1$
barbels appear at stage 44 and disappear at stage 61

Table 2. Some ontology rules in the *Xenopus* ontology

During the design of our ontology, we first tried to establish a list, as exhaustive as possible, of rules that are likely to have impact on the pictures, for example:

- when an element was tagged as *marked* with the *SBA lectin*, it is not possible to see *zone 9* of the *principal olfactory bulb*;
- *barbels* are only present when the *Xenopus* stage is between 44 and 61.

Furthermore, we carried out the same process for rules that were not expected to be related to the content of the pictures:

- a *Xenopus* can only be in one of the following states at any particular time: *embryo, tadpole* or *adult*;
- since the metamorphosis phases don't have strict temporal boundaries, either of two adjacent *stages* might be accepted for a *Xenopus*.

These rules lead us to finding out which classes are disjoint from each other (Figure 11 shows a disjointness that is related to the domain knowledge), and which classes could be used for restricting annotation fields (e.g., an adult *Xenopus* can only be in a *stage* above 61) or for defining patterns.

---

[4] Due to these nodes' positions: **Stage_d_evolution** → **Stage 47** and **Xenopus** → **Tete** → **Cerveau** → **Buble Olfactif**, then **zone 8**

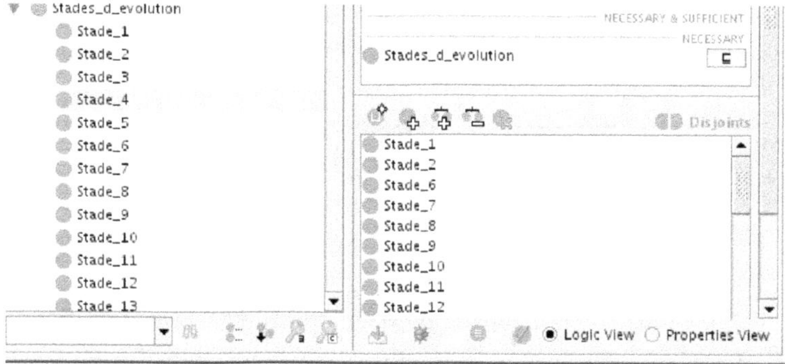

Fig. 11. Biological evolution is a continuous process, thus it is not always possible to strictly discriminate between two stages of a given individual. A *Xenopus* in stage 4 cannot be in a stage other than 3, 4, or 5

Analogously to the work described by [26], we have gone through the initial phase of the ontology development. We have gone through several refinement and evaluation cycles in order to be sure that the ontology reflects biologists' domain knowledge.

2.5 An error decision matrix

Error recovery is an important issue if users with different levels of expertise participate in the annotation process. We chose to define an error decision matrix in order to guide the annotation process. Since such a matrix must be complete (to take into account all possible cases), we based it on Missikof's [20] classification of annotation errors when using an ontology. We define a protocol for error recovery in terms of an Error Decision Matrix that indicates the corresponding action to be taken on the ontology for each kind of annotation error and each kind of user. Each row of the Error Decision Matrix is expressed in terms of a UML activity diagram to be used by the domain technicians. Table 3 shows the decision matrix as of this writing. It uses keywords from [5] to avoid ambiguity in actions. When instantiating an application, it is helpful to minimize future annotation problems by providing an associated correction mechanism for each annotation problem.

This decision matrix makes it possible to define a protocol for each group of users (domain technicians, domain experts, and specialists of knowledge representation) for each kind of annotation errors. Each protocol starts by an action by a domain technician. Depending on the kind of error, the occurrence of the problem and its possible solution are sent to the appropriate kind of user. For example, the *Name Path* error resolution has to be handled by a specialist of knowledge representation (Figure 12), and *Coverage* error correction has to be handled by either a domain expert or a specialist of knowledge representation (Figure 12).

Table 3: Error Decision Matrix describes necessary or mandatory modification operations on the ontology when an annotation error is detected.

Error type (from [20])	Example	Domain technicians	Domain experts	Specialists of knowledge representation
A Name path (different labels for the same content)	*Zone 8 of olfactory bulb* is the *accessory olfactory bulb*	MAY warn domain expert	MAY warn specialists of knowledge rep.	MUST check ontology and adjust if needed
B Encoding (different formats of date or units of measure)	Size of an element expressed in *number of pixel and resolution and scale;* or in *millimeters*	MAY modify content	MUST verify if both units are pertinent	SHOULD choose an encoding
C Structure (different structures for the same content)	A composed image process referred to as a string or as the list of all processes composing it	MUST NOT do anything	MUST decide if both structures are pertinent	SHOULD choose a structure
D Content (different content denoted by the same concept - typically expressed by enumeration)	*The Xenopus* concept described by different enumeration items	MAY modify content	SHOULD modify content	SHOULD NOT do anything

Error type (from [20])	Example	Domain technicians	Domain experts	Specialists of knowledge representation
E Coverage (presence/absence of information: information is mandatory for end-users but was not included in ontology)	Brand new *staining* not included in ontology	MUST NOT do anything	MUST warn a knowledge specialist about the lack of information / MAY add information	MUST adjust ontology / MAY suggest better ontology structure
F Precision (the accuracy of information)	Size of an element expressed with a semantical annotation such as "big" or with a number	MUST NOT do anything	MUST give a knowledge specialist the required precision	MUST adjust ontology
G Abstraction (level of specialization refinement of the information)	Annotation referring to *olfactory bulb* instead of *principal olfactory bulb*	SHOULD adjust with full precision	MUST adjust with full precision	MUST NOT do anything
H Granularity (level of decomposition refinement of the information)	A *Xenopus* head represented as a whole or as an aggregation of *barbels, olfactory bulb,* etc.	SHOULD adjust with full precision	MUST adjust with full precision	MUST NOT do anything

3 Beyond the *Xenopus* application

In this section, we present the second phase of our work towards the use of a domain ontology for image database engineering. At the end of the section, we outline our perspective on the metamodeling field.

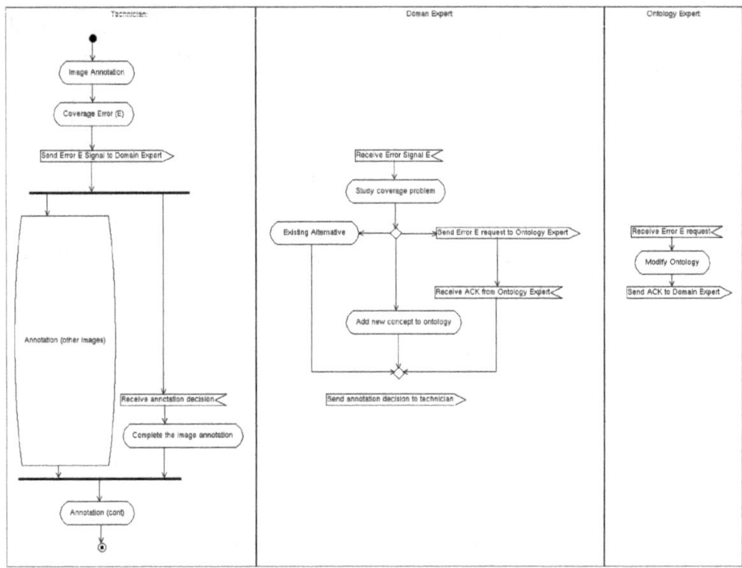

Fig. 12. Protocol in case of error E: describes necessary actions for each kind of user when a *domain technician* **detects** a *Coverage* error in the ontology

3.1 Summary of ontology issues

As stated in Section 1, domain ontologies can be used to define attribute ranges in a model in terms of ontology sub-hierarchies. For example, the sequence of the stage attributes of a *Xenopus* description is located at the **Evolution_stages** node of the *Xenopus* ontology. In this section, we present some complementary mechanisms:

Semantic consistency check. An environment for ontology design generally offers interesting tools called *reasoners* for dealing with semantic consistency. For example, W3C proposed OWL, an ontology language for Semantic Web in [2, 32]. One sub-language of OWL, OWL-DL (DL standing for *Description Logic*) can store rules and relations of a domain, and OWL-DL descriptions can be processed by reasoners like RacerPro [12, 16, 29]. In our application, we can use reasoners in two particularly interesting ways.

First, a reasoner can detect annotation errors. For example, if we assume the following rules:
- *embryo* and *adult* are disjoint classes,
- *stage 3* and *stage 47* are disjoint classes, *Xenopus* in stages between 1 and 28 is an embryo,
- *Xenopus* in stages higher than 66 is an adult.

Then, when a technician annotates an image with *adult* attribute, the reasoner will notice that possible *stages* must be at least 66; later, when the technician tries to annotate the image with *stage 3*, the reasoner will notice that this second annotation is nearly equivalent to *embryo* and will point out this inconsistency between *adult* and *embryo* to an expert. It is particularly important to use a reasoner to check consistency of the whole database when the domain knowledge evolves (thus implying an ontology modifications). For example, the *Xenopus* ontology initially contains an element called *barbel*. It is also established that such barbels appear at stage 44 and disappear at stage 61. The modified ontology includes the derived constraints and it will be thus necessary to check them for potential inconsistencies.

Second, a reasoner is also able to evaluate all the rules and make inferences from annotations in order to propose an inferred classification. For example, if a *Xenopus* is annotated with *stage 3*, the reasoner will propose that this *Xenopus* is also a member of the *embryo* class.

Image annotation guidelines. A domain ontology can provide worthy help with annotating images since the ontology's structure enforces the domain knowledge organization on annotations (in our *Xenopus* database, *olfactory bulb* cannot be annotated as something other than a part of the *Xenopus* brain). Furthermore, the ontology's rules can be used as a very valuable framework for annotations. For example, when the first rule of Table 2 is applied (barbels appear at stage 44 and disappear at stage 61), we notice the following:

- when the user selects a stage between 44 and 61, it is possible to annotate an image with the *barbel* concept. When the stage is lower than 44 or higher than 61, it is not possible to annotate with the *barbel* concept.

- when the user annotates an image with the *barbel* concept, it implies that the stage lies necessarily between 44 an 61. Combining two ontology relations of Table 2 (i.e., "*barbels* appear at stage 44 and disappear at stage 61" and "*Xenopus* in a *stage* between 28 and 66 is a tadpole"), it is possible to infer that when an annotation is made with *barbel*, then the corresponding *Xenopus* is a tadpole.

Physical extraction. Although many algorithms exist for image processing, it might not always be possible to carry out the derived computations on images: different reasons (such as processing power, volume of the information storage, etc.) prevent us from running all the available tools on all the pictures of a database.

Similarly to our use of an ontology in assisting with semantical annotations, the same ontology can be used as a guideline for determining relevant physical extractors. We propose to make an inventory of pertinence among physical extractors and annotations (seen as a single annotation or as a set of annotations) using three relations:

- Extractor A *is meaningful* for Annotation B (B can be a set of annotations): when an image is annotated with B, it is meaningful to perform computations using extractor A.

- Extractor A *might be meaningful* for Annotation B: when an image is annotated with B, it might be meaningful to perform computations using extractor A. In such a case, it is possible to let a person decide whether to perform the computation or instead to wait until the relation can be improved as the domain knowledge evolves.
- Extractor A *is useless* for Annotation B: annotation B implies that it is useless to perform computations using A.

For example, if a *Xenopus* image is annotated with "thermal camera", every real-colour-based extractor can be considered irrelevant and thus useless but an extractor that will compute temperature from an image would be meaningful.

Such a mechanism can use global semantical annotations on images or local annotations on image parts (by using a strategy which alternates phases of physical and semantical annotations).

3.2 Summary of metamodeling issues

In this chapter we have described our work on the *Xenopus* application. This work aims at establishing a coupling between an image database model and the database application, particularly its interface for image annotation (extraction of both physical and semantical attributes). The current work on the *Xenopus* application is a part of the more general work on coupling metamodeling architectures and ontologies. As discussed in the above section, an instance-level coupling is used for image annotation. A model-level coupling is achieved by attribute ranges and pattern descriptions that are defined in terms of ontology classes. As stated in [28], a third level of coupling between an ontology and metamodel levels can be achieved. For example, the stereotype relations ≪Rel≫ with attribute type set to *spatial* can be associated with spatial relations in the BFO-part[25] of the domain ontology.

Our upcoming work on the ontology will thus reuse knowledge from existing biomedical ontologies, such as relations from the Open Biomedical Ontologies (OBO [24, 9]) at three levels of our architecture. "Redesigning the wheel" can be avoided in those cases that have already been dealt with by the reused ontology. Discussions between different ontologies community will be easier and less ambiguous if everyone agrees on using the same concepts in the same representation.

References

1. K. Baclawski, M.Kokar, P. Kogut, L. Hart, J. Smith, W. Holmes, J. Letkowski, and M. Aronson. Extending UML to Support Ontology Engineering for the Semantic Web. In *Proceedings of the International Conference on UML, UML'01, Toronto, Canada*, 2001.
2. Tim Berners-Lee. Semantic web - xml2000, 2000. `http://www.w3.org/2000/Talks/1206-xml2k-tbl/`.

3. Philip A. Bernstein. Applying model management to classical meta data problems. In *Proc. of the First Biennial Conference on Innovative Data Systems Research (CIDR03), USA*, 2003.

4. Laurent Besson. *IkoSem : une plateforme pour l'ingénierie des bases de données images*. PhD thesis, Université de Bourgogne, 2004.

5. Scott Bradner. Key words for use in rfcs to indicate requirement levels, 1997. `http://www.ietf.org/rfc/rfc2119.txt`.

6. S. Brockmans, R.M. Colomb, E.F. Kendall, E.K. Wallace, C. Welty, and G. Tong Xie. A model driven approach for building owl dl and owl full ontologies. In *Proceedings of the 5th International Semantic Web Conference (ISWC'06)*, 2006.

7. Walter W. Chang. A discussion of the relationship between RDF-schema and UML. W3C note, W3C, August 1998. http://www.w3.org/TR/1998/NOTE-rdf-uml-19980804.

8. S. Cranefield and M. Purvis. Uml as an ontology modelling language, 1999.

9. OBO Foundry. Open biomedical ontologies (obo). `http://www.obofoundry.org/`.

10. Frédéric FRST. L'ingénierie ontologique. Technical report, Institut de recherche en Informatique de Nantes, 2002.

11. Arnaud Gaudin and Jean Gascuel. 3d atlas describing the ontogenic evolution of the primary olfactory projections in the olfactory bulb of xenopus laevis. *The Journal Of Comparative Neurology*, 2005.

12. Racer Systems GmbH. Racerpro : Owl reasoner and inference server for the semantic web, 2004. `http://www.racer-systems.com/`.

13. Matthew Horridge. A practical guide to building owl ontologies using the protégé-owl and co-ode tools, 2004. `http://protege.stanford.edu/`.

14. Paul Kogut, Stephen Cranefield, Lewis Hart, Mark Dutra, Kenneth Baclawski, Mieczyslaw Kokar, and Jeffrey Smith. Uml for ontology development. *Knowl. Eng. Rev.*, 17(1):61–64, 2002.

15. C.C. Leung, C.H. Chan, F.H.Y. Chan, and W.K. Tsui. B-spline snakes in two stages. In *Proceedings of the 17th International Conference on Pattern Recognition*, 2004.

16. Deborah L. McGuinness and Frank van Harmelen. Owl web ontology language, 2004. `http://www.w3.org/TR/owl-features/`.

17. Melnik. Representing uml in rdf, 2000.

18. S. Melnik, E. Rahm, and P. A. Bernstein. Rondo: A programming platform for model management. In *Proc. of the ACM SIGMOD 2003, USA*, 2003.

19. Tim Menzies. Cost benefits of ontologies. *Intelligence*, 10(3):26–32, 1999.

20. M. Missikof, F. Schiapelli, and F. Taglino. A controlled language for semantic annotation and interoperability in e-business applications.

21. The Gene Ontology. the gene ontology editorial style guide, 2007. `http://www.geneontology.org/GO.usage.shtml`.

22. J.Z. Pan and I. Horrocks. Metamodeling architecture of web ontology languages. In *SWWS*, pages 131–149, 2001.

23. A. Th. (Guus) SCHREIBER, Barbara DUBBELDAM, Jan WIELEMAKERr, and Bob WIELINGA. Ontology-based photo annotation. *IEEE Intelligent Systems*, 16(3):66–74, 2001.

24. Barry Smith, Werner Ceusters, Bert Klagges, Jacob Kohler, Anand Kumar, Jane Lomax, Chris Mungall, Fabian Neuhaus, Alan Rector, and Cornelius Rosse. Relations in biomedical ontologies. *Genome Biology*, 6(5):R46, 2005.

25. Andrew D. Spear. Ontology for the twenty first century: An introduction with recommendations, 2006.

26. York Sure, Steffen Staab, and Rudi Studer. Methodology for development and employment of ontology based knowledge management applications. *SIGMOD Rec.*, 31(4):18–23, 2002.

27. William Swartout and Austin Tate. Guest editors' introduction: Ontologies. *IEEE Intelligent Systems*, 14(1):18–19, 1999.

28. Marie-Noelle Terrasse, Marinette Savonnet, Eric Leclercq, George Becker, and Thierry Grison. Do We Need Metamodels AND Ontologies for Engineering Platforms? In *Proceedings of the 1st ICSE Int. Workshop on Global Integrated Model Management, China*, 2006. URL http://portal.acm.org/citation.cfm?id=1138304.1138310.

29. The National Center For Biomedical Ontology. Protege, 2006. `http://protege.stanford.edu/`.

30. *OMG Unified Modeling Language Specification*, March 2000. Version 1.3, Available at URL http://www.omg.org.

31. Mike Uschold and Michael Grüninger. Ontologies: principles, methods, and applications. *Knowledge Engineering Review*, 11(2):93–155, 1996.

32. W3C. Semantic web, 2004. `http://www.w3.org/2001/sw/`.

33. Consortium W3C. Intellectual rights faq, 2004. `http://www.w3.org/Consortium/Legal/IPR-FAQ-20000620`.

34. Nianbin Wang and Xiaofei Xu. A method to build ontology. In *HPC '00: Proceedings of The Fourth International Conference on High-Performance Computing in the Asia-Pacific Region-Volume 2*, page 672, Washington, DC, USA, 2000. IEEE Computer Society.

35. B. J. Wielinga, A. Th. Schreiber, J. Wielemaker, and J. A. C. Sandberg. From thesaurus to ontology. In *K-CAP '01: Proceedings of the 1st international conference on Knowledge capture*, pages 194–201, New York, NY, USA, 2001. ACM Press.

User-aware adaptation by subjective metadata and inferred implicit descriptors

Cezar Pleşca, Vincent Charvillat and Romulus Grigoraş

IRIT-ENSEEIHT, Toulouse, France. E-mail: `plesca,charvi,grig@enseeiht.fr`

Summary. Universal Media Access faces a wide range of dynamic execution environments with varying networks, terminals, user profiles. In order to handle these highly dynamic contexts, we introduced a formal approach using adaptation agents. Multimedia adaptation agents perceive the context states thanks to observations and carry out adaptation actions. The decision-making principle is based on reinforcement learning and uses Markov Decision Processes. We have already applied this framework to several case studies: adaptive streaming, adaptive content delivery, user profiling etc. Based on our recent work on agent-driven adaptation of multimedia content, this chapter deals with appropriate metadata support for our adaptation framework. Although compatible with MPEG-21, our approach requires changes in the management of metadata. The adaptation agent itself is metadata. Our context descriptors are either observable or only partially observable. The final point is that our decision models naturally handle subjective metadata such as user interest. As a result, we are able to abstract user interest descriptors that are implicitly generated by observing user interactions.

Keywords: multimedia adaptation, reinforcement learning, Markov Decision Process, implicit feedback, adaptation agent, embedded metadata

1 Introduction

The Universal Multimedia Access (UMA) initiative is coming from a phenomenal increase: i) in the quantity and complexity of content (including rich media content); ii) in the number of different users with different preferences; iii) in the heterogeneity of devices, digital storage and networks.

Far from the previous Internet or TV paradigms, the emerging use cases are further complicated by the inherently dynamic nature of mobile communication and interactive services. Users and providers alike have to deal with rapidly varying resources (bandwidth, load on terminals, etc.). There is, therefore, more need than ever before for context aware services and applications which are able to adapt dynamically to changing service provision conditions.

C. Pleşca et al.: *User-aware adaptation by subjective metadata and inferred implicit descriptors*, Studies in Computational Intelligence (SCI) **101**, 127–147 (2008)
www.springerlink.com

The description or modeling of such a "usage context" is a major issue of current research in adaptation within several research communities. Sometimes limited to content transcoding in the past, resource-based adaptation frameworks now integrate user profiles and preferences (e.g. the Usage Environment Description of MPEG-21 DIA [17]).

This work builds upon the formal adaptation framework that we have successfully used in our research. Our approach provides adaptation agents able to dynamically choose the best adaptation actions in a closed-loop manner.

Although compatible with MPEG-21 and in particular with ADTE interfaces, our approach requires changes in the management of metadata. We consider not only completely observable descriptors but also partially observable context variables. In our formal approach, a decisional agent manages these metadata by memorizing fully-observable contextual metadata. Therefore, **the agent can be embedded in the content and can be considered metadata itself.** We will show the significance of these changes and also the integration with resource-based adaptation agents.

The chapter is organized as follows. The Section 2 briefly introduces our metadata-based adaptation approach. Next, we describe a case study and analyze different context descriptors useful for adaptation purposes. Following this analysis, we show the need of considering semantic descriptors and propose a classification of various metadata from a novel perspective in Section 2.3. Then, we introduce our general approach based on reinforcement learning in Section 3. The fourth section describes Markov Decision Process (MDP) formalism along with its extension for partially observable contexts (POMDP). We make use of our case study to show how these models can be rendered using the appropriate context descriptors. Finally, experimental results are presented before summarizing our metadata adaptation recipe.

2 Metadata for multimedia adaptation

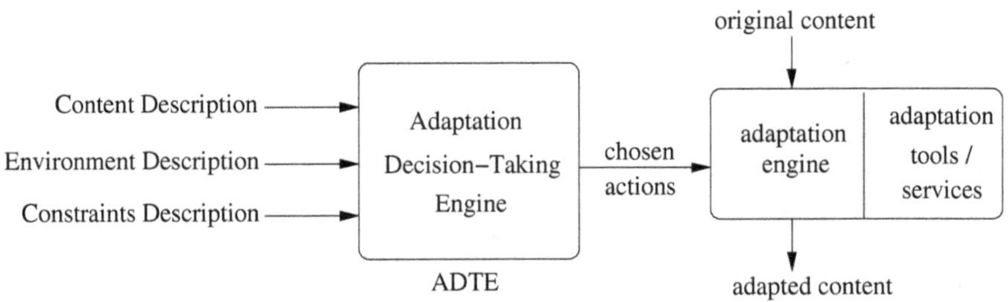

Fig. 1. Decision-taking engine

In order to handle the heterogeneity of terminals and networks and the diversity of user preferences, adaptation of multimedia resources is necessary.

In context-aware adaptation any "one-size fits all" approach is conceptually and technologically unrealistic. Hence researchers try to define common description toolkits. These descriptors or metadata enable both open proposals for adaptation and interoperability between adaptation components or services. Such metadata driven techniques propose to associate the content with metadata that provides the relationship between possible adaptation actions and various resources' characteristics.

2.1 Related work

Within the W3C framework, a CC/PP profile (Composite Capabilities / Preferences Profiles) is a description of device capabilities and user preferences that can be used to guide the content adaptation [8]. A CC/PP profile contains a number of attributes and their associated values that allow determining the most appropriate form of a resource to be delivered.

Within the ISO family, MPEG-21 aims at defining a multimedia framework for transparent and augmented use of multimedia resources across a wide range of networks and devices used by different communities. The fundamental unit of distribution and transaction within MPEG-21 is the Digital Item (DI). A Digital Item is a combination of resources (e.g. videos, audio tracks or images), metadata (i.e. descriptors and identifiers) and structure for describing the relationships between the resources.

MPEG-21 Digital Item Adaptation [17] specifies the syntax and semantics of tools that may be used to assist the adaptation of Digital Items, i.e., the Digital Item Declaration and resources referenced by the declaration.

Describing the context is necessary but not sufficient for deciding how to perform adaptation. The decision taking is a key component for context-aware adaptation. Picking up a feasible solution is very different than selecting the optimal adaptation decision. Figure 1 introduces a possible architecture for adapting the content in an optimal way. Following several MPEG-21 based works [5, 9], we emphasize the distinction between the **adaptation decision-taking engine** and the **adaptation engine**.

The adaptation decision-taking engine (ADTE) takes several context descriptors as inputs and outputs a chosen adaptation action or decision. Firstly, the input comes from the user side and is a declarative description of its usage environment (i.e. device capabilities, network conditions and user preferences). This description is contained in a Usage Environment Description (UED) whose specification is part of the MPEG-21 Digital Item Adaptation (DIA) standard [17]. Secondly, descriptions of the content and the available adaptation tools are also provided. The content descriptions are expressed in terms of MPEG-7 media information metadata and contain information such as media encoding, color or spatial resolution.

The chosen action is then forwarded to the adaptation engine which executes the required transformation steps on the original resource according to the adaptation decision and produces the adapted multimedia resource.

We build upon an original learning-based approach to dynamic adaptation introduced in [3]. The context-based adaptation is addressed by integrating human factors and available resources. We specifically handle highly dynamic contexts where available resources are limited and time-varying and where users' interactions are also uncertain. Such highly dynamic and incertain contexts are illustrated through the case study described in the next section: a movie presentation system on mobile terminals.

2.2 Case study: a movie browsing service

Figure 2 introduces an information system accessible from mobile terminals such as PDAs. A keyword search allows the user to obtain an ordered list of links to various movie descriptions. Among this list, the user can follow a link towards an interesting movie (the associated interaction will be referred to as *clickMovie*); then, he or she can consult details regarding the movie in question. This consultation will call on a full screen interactive presentation. Having browsed the details for one movie, the user is able to come back to the list of query results (interaction *back* on the Figure 2). It is then possible to access the description of a second interesting film. The index of the accessed movie description will be referred to as n.

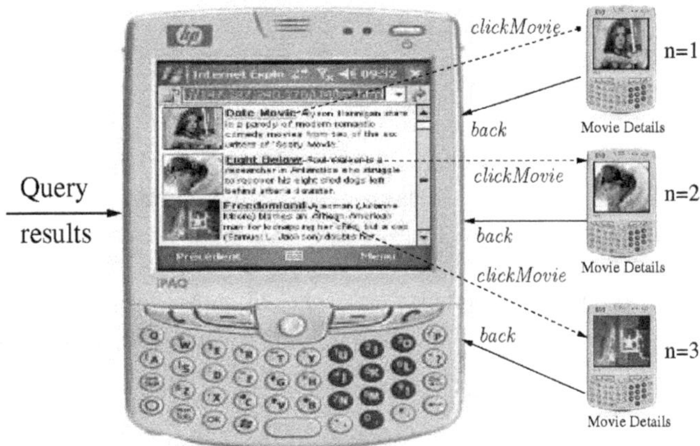

Fig. 2. Information system of movie descriptions

To simplify the context modeling, we choose to consider the browsing sequence indexed by n. Our problem now becomes one that aims at adapting the content (movie description) presented during this sequence. The execution environment is dynamic because of the bandwidth's (bw) variability in a wireless network.

To present the details of a movie, three descriptions are possible (Figure 3). The poor "Textual" version (referred as T) groups together a small poster image, a short text description and links pointing to the production photos as well as a link to the video trailer. The intermediary version (I) provides a slideshow of production photos and a link to the trailer. The richest version (V) includes, in addition, the video trailer.

Fig. 3. Basic (T), intermediary (I) and rich (V) versions of movie details

For this scenario, adapting the content could be formulated as the problem of choosing one of the three versions (e.g. T, I or V). Such a decision has to be taken sequentially, for each movie description requested by the user. Hence, we are searching for the best sequence of decisions (versions) according to description index (i.e n), available resources (i.e. bw) and user interest. By their nature, these context variables are different with respect to their ambiguity and subjectivity. The following section proposes an analysis of these differences from a novel perspective.

2.3 Subjective semantic descriptors

This section presents an original approach for handling context variables or descriptors within our adaptation framework. The novelty comes from the distinction between observable and partially observable descriptors. This distinction is made by the designer of the decision-taking engine.

Our first idea is to classify context descriptors according to two semantic features: the **subjectivity** level and the **observability** level. For example, our case study uses 3 context variables (n, bw and It) which are very different with respect to these semantic features. Obviously n (the movie index) is an objective variable: no subjectivity or ambiguity when determining its value. Therefore n is an observable descriptor.

Conversely, It (the user interest level) is a highly subjective variable by nature and cannot be directly observable. Only some observations such as user interactions can provide hints on its hidden value. This inferrence process is called *implicit feedback* [7]. Hence It is a partially observable descriptor.

As for bw (the available bandwidth), it is clearly an objective variable. Nevertheless, measuring its value is notoriously difficult. There is a significant

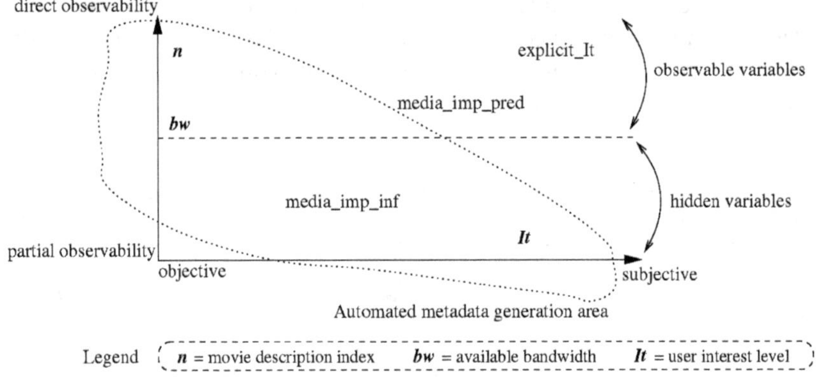

Fig. 4. Contextual variables

amount of uncertainty when estimating it from current network parameters (e.g. RTT, loss rate). On the observability axis, bw lays between n and It. Figure 4 places these three contextual variables with respect with both axis. In addition, we provide some extra examples of metadata.

Let's consider the composite page illustrated in Figure 5 whose content is originally designed to be played on a desktop device. To adapt this rich content to a mobile terminal, Mariam [1] proposes the use of semantic information descriptors such as *media importance*.

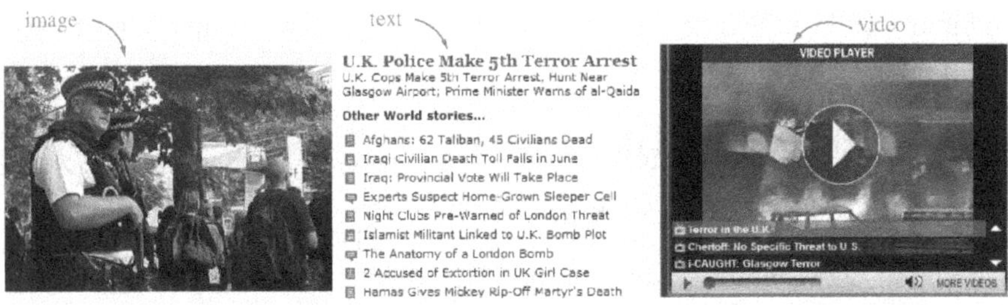

Fig. 5. News page composed of various media with different importance

The *importance* attribute describes the semantic importance of the corresponding media object represented on three levels: high, medium and low. For instance, the semantic information attributes for the textual component from the news page could be expressed in a XML-like description as follows: `<sid:Object id="text" ref="T" importance="high" role="key">`. While adaptation of *high* and *medium* importance objects are relatively privileged, *low* importance media objects may be omitted from the page in the case of limited available resources.

If the importance level is predefined by the author (`media_imp_pred`), the corresponding descriptor is probably neither fully objective nor as subjective as the user interest. If the media importance level is automatically inferred (the descriptor is called `media_imp_inf`), its observability becomes more difficult to achieve. Nevertheless, this descriptor is often obtained from the interactions with the content and it is therefore less subjective than the one presumed by the author. Similarly, if the user interest level is obtained by explicit feedback, the associated descriptor `explicit_It` tends to be fully observable.

As a consequence of this analysis, the figure 4 shows the area for which metadata can be automatically generated/inferred. The more subjective or the less observable the descriptors, the more difficult the inference. This can help the designer of an ADTE for choosing the observability threshold beyond which the descriptors should be considered as hidden.

3 Decision-taking through Reinforcement Learning

In order to design ADTE, many authors propose to use artificial learning techniques to select the right decision and/or the appropriate implementation of adaptation mechanisms (see [12] for a review or [5] for an relevant example). In this case, a description of the running context is given as input to a decision-making agent that predicts the best adaptation actions according to what it has previously learned.

We extend this idea in line with a reinforcement learning principle [3]. Similar approaches were proposed for recommandation systems [13]. Nevertheless, taking into account subjective context variables to drive adaptation decisions has, to the best of our knowledge, never been explored.

3.1 A closed-loop adaptation agent

In the context of reinforcement learning (RL)[15], let us consider an adaptation agent and its context. This context integrates the user, his mobile terminal, the network(s), the content (documents, media etc.). The adaptation agent can perceive (at least partially) the characteristics of the context. These perceived characteristics give the state of the agent.

The agent can influence the context: it can transform media, documents, choose the media to be displayed, take actions on a user's behalf, relocate content, make bandwidth reservations, make a combination of all these actions or do nothing. Among various possible actions it can take in a given state, an agent has to find the best one. The agent's behaviour is called "decisional policy", that is simply a function that returns an action for each state.

The agent learns from its interaction with the context: as a reaction to its decisions, the context returns observations and a reward (Figure 6). The observations allow the agent to estimate the current state. The aim is to operate a "trial and error" type of learning, located half-way between supervised and unsupervised learning. The agent tries to reinforce the decisions that resulted in a good accumulation of rewards and, conversely, tries to avoid unfruitful decisions. This implies an iterative update of its "decisional policy" that should ideally tend to an optimal policy. As the context states can be described as Usage Environment Descriptions (see Section 2.1), the optimal adaptation policies can be easily expressed following MPEG-21 Digital Item Adaptation (DIA) standard. This makes our adaptation approach compatible with ADTE interfaces from MPEG-21 general adaptation framework.

The main difficulties of this general approach are the description of the context and the identification of "rewards", that is performance criteria. It is tricky to find states that are expressive enough while avoiding a combinatorial explosion. The current research in AI concentrate on these points.

3.2 Adapting the content through RL

The case study introduced in Section 2.2 allows us to illustrate the application of the general RL-based adaptation framework. Figure 7 provides an intuitive answer to the key question: what are the best adaptation decisions ? Our adaptation agent takes the current context as an input signal. The input should include the movie index n, the available resources (bandwidth bw) and eventually the user interest. The adaptation agent outputs a decision by prediction. This prediction is based on a training data set composed of previously observed resource limitations and user interactions.

There are still three remaining questions. How can the dynamics of the context be handled ? What are performance criteria for evaluating a decision ? How user interest can be inferred ?

Handling context dynamics

Dynamic contexts require the use of a closed-loop decision process since the decision at some point in time may influence the future events. For instance, if the adaptation strategy selects the richest version (V) this impacts the behavior of a user who experiences bad network conditions (low bandwidth).

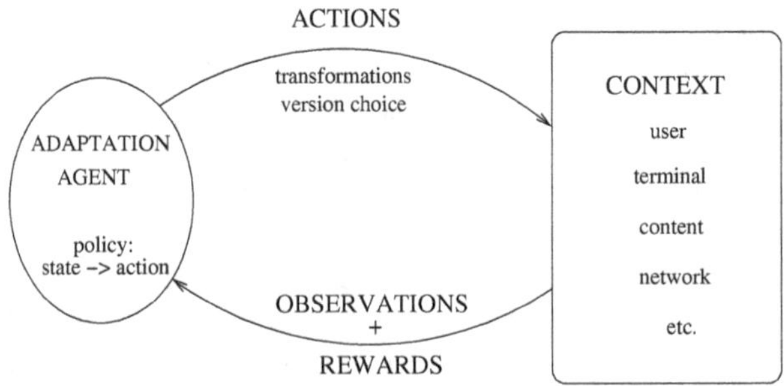

Fig. 6. Context-based adaptation agent

Although strong latencies can be tolerated while browsing the first query results (small index n), the latencies quickly become unacceptable with an increasing n, incurring the risk to have the user end the session.

Choosing performance criteria

A user-aware performance criterion can drive the decision. Figure 7 shows a closed-loop decision process in which the agent receives as input both the current context and a performance reward. For our example, the agent receives a reward if the version selection (T, I or V) is not challenged by the user. The reassessment of a version T as being too simple is suggested, for example, by an *endImages* (EI) event associated with the full consumption of the pictures. In the same way, the reassessment of a version V as being too rich, is indicated by a partial consumption of the video.

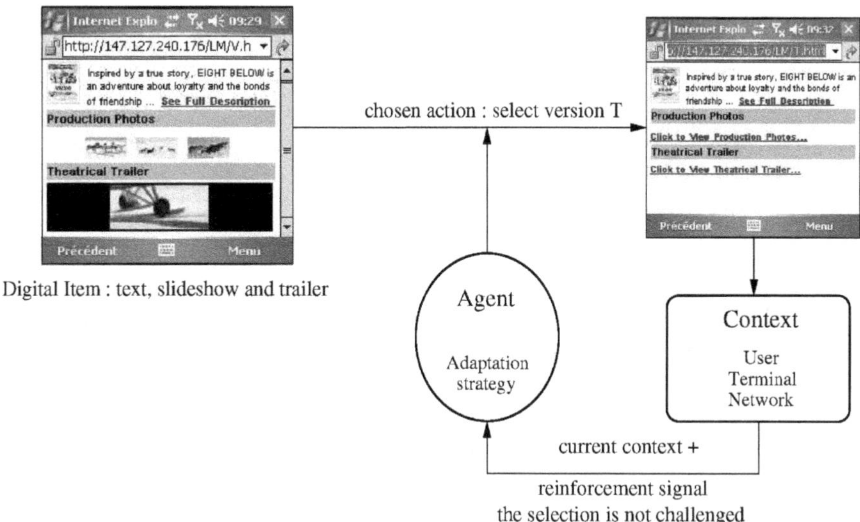

Fig. 7. Selecting the content version on a mobile terminal

Therefore, the agent uses a "trial and error" strategy that converges to an appropriate adaptation policy. Three properties of a good adaptation policy can be identified:

1. The version chosen for presenting the content must be simplified if the available bandwidth bw decreases [1].
2. The version must be simplified if n increases: it is straightforward to choose rich versions for the first browsed movie descriptions that are probably the most pertinent ones [2].

---

[1] T (text) is simpler than I (T & images), itself simpler than V (I & video)
[2] we should avoid large latencies for big values of n and small bw

3. The version must be enriched if the user shows a high interest for the query results. The underlying idea is that a very interested user is more likely to be patient and tolerate more easily large downloading latencies.

Inferring user interest

There are two main methods for estimating user interest. Explicit feedback techniques (Figure 8) use modal dialog windows to directly ask the user his/her interest level. Conversely, Implicit feedback is experiencing a growing interest, since it avoids bringing together significant collections of explicit returns (which is intrusive and expensive) [7]. These IF methods are used in particular to decode user reactions in information search systems [6]. Among the studied implicit feedback signals one can consider are: the total browsing time, the number of clicks, the scrolling interactions and some characteristic sequences of interactions. In our work, user interest is estimated using IF by interpreting interaction sequences.

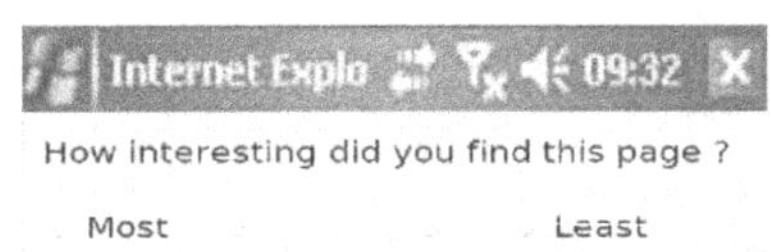

Fig. 8. Explicit feedback to observe user interest

Intuitively, there are two classes of interactions according to what they suggest: either an increasing interest (following links to see the media: *startSlide, play*) or a decreasing interest (*stopSlide, stopVideo, stopDownload*). Thanks to this example, it is now straightforward to argue the use of reinforcement learning methods for handling dynamic adaptation.

3.3 Modelling hidden contexts states

When describing context states, both hidden and observable variables can be used. A state becomes hidden if it contains at least one hidden descriptor. For instance, Figure 9 shows three hidden states that only differ according to user interest levels, ranging from Small to Big.

Following the work of [16], it is natural to model the sequence of events or observations (i.e. user interactions) produced by a Hidden Markov Model (HMM) (see [4] for further details). Given a sequence of observations, an HMM can provide the agent with the most likely underlying sequence of hidden states or the most likely running hidden state.

Second row of Figure 9 shows that the three probability distributions of the interactions are different for each state. Obviously, if the probability of stop events is high, it suggests (or reinforces) a smaller (or small) interest. Despite these hints, it is generally impossible to fully identify a hidden context

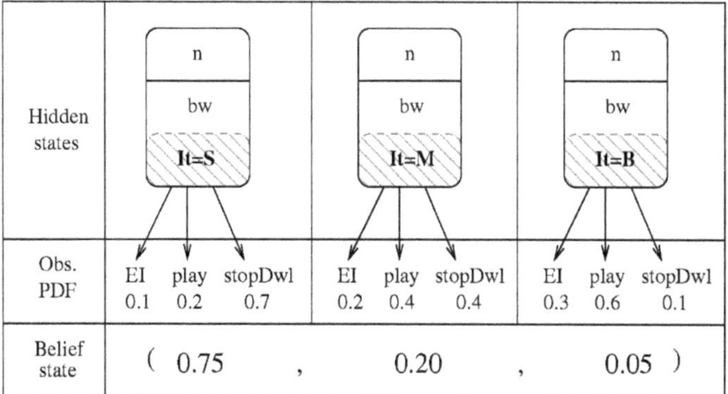

Fig. 9. Hidden states, their observations probability distributions and the belief state

state. In order to handle this uncertainty, belief states provide an efficient representation. Third row of our figure shows such a belief state which is simply a distribution probability on the set of possible hidden states. This belief state suggests that the most likely user interest level is small.

4 MDP-based adaptation models

Our dynamic adaptation approach fits naturally into frameworks of *sequential decisional policies under uncertainty*. In these frameworks, the uncertainty comes from two sources. On the one hand, the dynamic of the context can be random as a consequence of available resources' variability (for example the bandwidth); on the other hand, the effect of an agent's decision can be itself random. For example, if an adaptation action aims to anticipate user interactions, the prediction quality is obviously uncertain and subject to the user's behavior variations.

In this situation, by adopting a Markov definition of the context state, the agent's dynamics can be modeled as a Markov Decision Process (MDP). This section introduces this formalism. We initially assume that context state variables are observable by the agent which makes it a sufficient condition to identify the decision state without any ambiguity.

Then we take a step forward by refining adaptation policies according to user interest. We estimate sequentially this hidden information through user behavior as suggested by research on the evaluation of "implicit feedback". Therefore, the new decision-making state contains at the same time observable variables as well as a hidden element associated with user interest. We then move on from a MDP to a POMDP (Partially Observable Markov Decision Process) that allows us to handle partially observable contexts such as the one presented in our case study.

5 Decision-taking models

In the previous section, we have shown that reinforcement learning is a relevant framework to tackle dynamic adaptation problems. In this section, we introduce the formalization of reinforcement learning with Markov Decision Processes (MDP) whose states are completely or partially observable.

5.1 The Markov property and MDP definition

In the reinforcement learning (RL) framework the agent makes its decision by applying a policy. A policy is a function that takes a signal from the environment as input and outputs a decided action. Figure 6 shows such a function of the environment's state in the RL closed-loop.

Since the decision are taken sequentially, we would like an input signal that summarizes past events compactly and retains only relevant information for decision-taking. Usually this requires more than instantaneous measurements but never more than the complete history of visited states.

According to Sutton, "a state signal that succeeds in retaining all relevant information has the Markov property" (see also the formal definition in [15] p.63). If an environment has this property, it has a one-step dynamics: the next state and the next reward can be predicted only from the current state and the current action. A reinforcement learning process that satisfies the Markov Property is called MDP (Markov Decision Process).

A MDP is a stochastic controlled process that assigns rewards to transitions between states [11]. It is defined as a quintuple $(S; A; T; p_t; r_t)$ where S is the state space, A is the action space, T is the discrete temporal axis of instants when actions are taken, $p_t()$ are the probability distributions of the transitions between states and $r_t()$ is a function of reward on the transitions. We rediscover in a formal way the ingredients necessary to understand the figure 6 : at each instant T, the agent observe his state $\sigma \in S$, apply on the system the action $a \in A$ that brings the system (randomly, according to $p(\sigma'|\sigma, a)$) to a new state σ', and receives a reward $r_t(\sigma, a)$.

As previously mentioned, we are looking for the best policy with respect to the accumulated rewards. A policy is a function π that associates an action $a \in A$ with each state $\sigma \in S$. Our aim is to find the best one: π^*.

The principle of the *Q-learning* algorithm (figure 10) is the following: after each observed transition $(\sigma_n, a_n, \sigma_{n+1}, r_n)$ the current value function Q_n for the couple (σ_n, a_n) is updated, where σ_n represents the current state, a_n the chosen action, σ_{n+1} the resulted state, r_n the immediate reward and R_n the accumulated reward for this experience. The updating formula trades off previous estimation of Q_n and accumulated reward R_n for the current experience. A reader interested in further details is referred to [10].

5.2 Partial Observation and POMDP Definition

In many cases, the observations captured by an agent (figure 6) are only partial and do not allow the identification of the context state without ambiguity. Therefore a new class of problems needs to be solved: Partially Observable Markov Decision Processes.

The states of the underlying MDP are hidden and only the observation process helps to rediscover the running state of the process. A Partially Observable Markov Decision Process (POMDP) is defined by an underlying MDP $(S; A; T; p; r_t)$ and a set of observations $\mathcal{O}$. Additionnaly, $O : S \rightarrow \Pi(\mathcal{O})$ is an observation function that maps every state s to a probability distribution on the observations' space. The probability to observe o knowing the agent's state s will be referred as: $O(s, o) = P(o_t = o | s_t = s)$.

Non-Markovian behavior. It is worth to note that, in this model, we loose a widely used property for the resolution of the MDPs, namely that the observation process is Markovian. The probability of the next observation o_{t+1} may depend not only on the current observation and action taken, but also on previous observations and actions.

Solving POMDPs is a much more difficult task than solving Markov Decision Processes (MDPs) [14] mainly because a given observation can be associated with many different states of the underlying MDP.

Stochastic policy. It has been proved that the results obtained for the V and Q convergence using MDP resolution algorithms are not applicable anymore. The POMDPs provides stochastic policies and not deterministic ones, as in the case of MDP [14].

5.3 Resolution

The POMDP classic methods attempt to bring back the resolution problem to the underlying MDP. Two situations are possible. If the MDP model is known, one can not determine the exact state of the system but a distribution probability on the set of the possible states (a *belief state* - Figure 9). In the second situation, without knowing the model parameters, the agent attempts to construct the MDP model relying only on observations' history.

Initialize Q_0
for $n = 0$ **to** $N_{\text{tot}} - 1$ **do**
    $\sigma_n =$`chooseState`, $a_n =$`chooseAction`
    $(\sigma_n', r_n) =$`simulate`(σ_n, a_n)
    $R_n = r_n + \gamma \max_b Q_n(\sigma_n', b)$ /* the current expectation reward R_n */
    $Q_{n+1}(\sigma_n, a_n) \leftarrow (1 - \alpha_n(\sigma_n, a_n))Q_n(\sigma_n, a_n) + \alpha_n(\sigma_n, a_n)R_n$
end for
return $Q_{N_{\text{tot}}}$

Fig. 10. The *Q-learning* algorithm

Our experimental test bed uses the resolution software package provided by Cassandra [2] that works in the potentially infinite continuous space of belief states using linear programming methods.

6 Proof of concept

This section presents the POMDP model used for adapting the movie browsing service described in Section 2.2. We build upon the reinforcement learning principle (Figure 7) enriched with the implicit feedback approach to estimate user interest (Figure 9).

Table 1 shows the principles of our POMDP model. In this chapter we prefer to present guidelines rather than "dry" formalisms. A reader interested in finding more about our POMDPs and MDP models can refer to the full description provided in [10].

6.1 POMDP model dynamics

First row of table 1 contains a snapshot of the user interactions sequence: selection of the n-th movie of the result list (Figure 2), navigation inside the description of this movie, return to the result list then selection of the $n+1$-th movie. This interaction sequence (i.e. $\{clickMovie, int_1, ...back, clickMovie\}$) is observed by the agent in the form of fully observable context descriptors. Similarly, we assume that the agent can be provided, at each user interaction, with an estimation of available bandwidth (i.e. $bw_0, bw_1, ...bw_p$). Obviously, the agent knows the movie index (i.e n).

As shown in the third row, the agent should be able to a) decide the best version v_n to submit to the user b) also decode the sequence of observations (user interactions) in order to infer user interest.

The agent applies a performance criterion, e.g. the sum of rewards or penalties observed along with the navigation. The rewarding system has already been suggested in Section 2.2. In brief, a version (T, I or V) is considered well chosen in a given context if it is not questioned by the user. The reassessment of a version T as being too simple is suggested, for example, by the full consumption of the pictures. In the same way, the reassessment of a version V as being too rich, is indicated by a partial consumption of the video.

The agent also includes a decision model (POMDP) that we can express using belief states such as those of Figure 9. Second column of table 1 contains a belief state with which the agent associates an optimal action v_n, obtained by reinforcement learning. This action depends not only on observable variables (n and bw), but also on probability distribution of user interest levels. In this example, the current interest level is most likely small. While the user browses a movie description, the agent takes no action. Nevertheless, along with interactions and bandwidth variations, the markovian dynamics of belief states updates the estimation of the current interest level.

What the user does	Selection of n-th film	Navigation inside the n-th film description					Return to the movie list	Selection of $n+1$-th film
What the agent	n	n						$n+1$
	bw_0	bw_1	bw_2	...	bw_k	bw_{k+1}	...	bw_{p-1}	bw_p
observes	$clickMovie$	int_1	int_2	...	int_k	int_{k+1}	...	$back$	$clickMovie$
	Subsequent rewards or penalties for agent decision							
What the agent does	Decide the version according to the estimated interest and the observations (n, bw)	• Interpret the sequence of interactions and bw variations according to the dynamics of the decision model. • Inferring user interest using Implicit Feedback (IF).						Next decision
POMDP Model 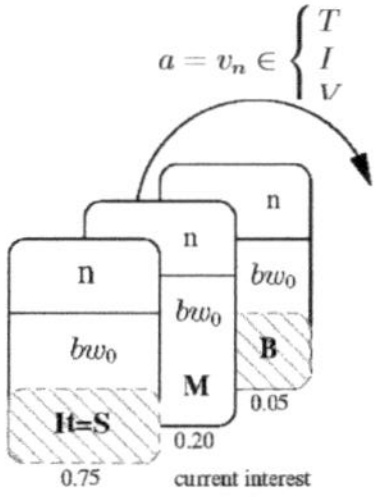 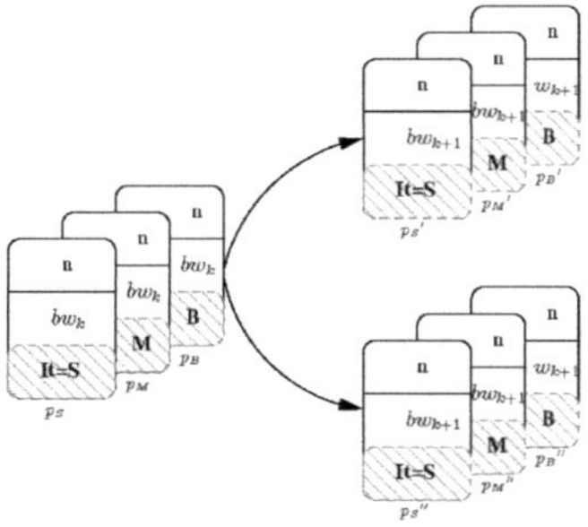 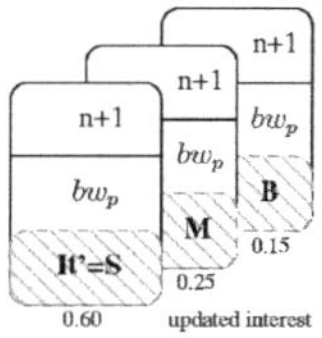								
Timeline								time →

Table 1. Outlines of the POMDP model used for movie browsing service adaptation

Cumulated rewards collected from the current movie description allow the learning agent to reinforce or not the choise of version v_n. For the next movie, the agent decides again the best version to be proposed according to its new belief state. As stated in Section 5.3, working in the continuous space of belief states gives a model with a potentially infinite number of states. This major drawback can be overcomed by transforming the model into a MDP.

6.2 Transforming the POMDP into a MDP

In order to solve the POMDP, we choose to transform it into a MDP following the second principle of POMDP resolution (i.e. for unknown models), as stated in Section 5.3. The estimation of each POMDP state is done by memorizing the sequence actions-observations between two successive decisional states.

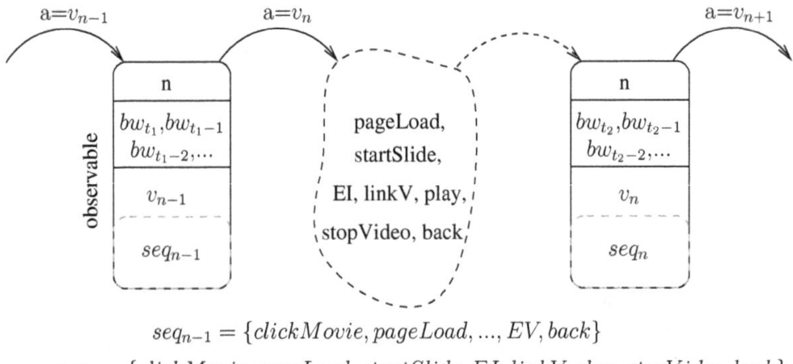

$$seq_{n-1} = \{clickMovie, pageLoad, ..., EV, back\}$$
$$seq_n = \{clickMovie, pageLoad, startSlide, EI, linkV, play, stopVideo, back\}$$

Fig. 11. MDP obtained from POMDP by memorizing past events

The resulting MDP state contains (Figure 11) both current observable elements (n and bw) and action-observation trajectory. The trajectory includes the action $a = v_n$ and the sequence of interactions $\{clickMovie, ..., back\}$. The previous bandwidth values ($bw_{-1}, bw_{-2}, ...$) are also taken into account.

6.3 Content adaptation policies

Experimental setup. Simulations are used in order to experimentally validate the proposed model. The developed software simulates navigations such as the one depicted in Figure 11 (i.e. the sequence of interactions seq_n). Every transition probability between two successive states of navigation (i.e. two consecutive user interactions) is a stochastic function of three parameters: bw, $It(n)$ and v. The bandwidth bw is simulated as a random variable uniformly distributed in a realistic interval. $It(n)$ represents a family of random variables, whose expectation decreases with n. The parameter v is the movie version proposed to the user. More details can be found in [10].

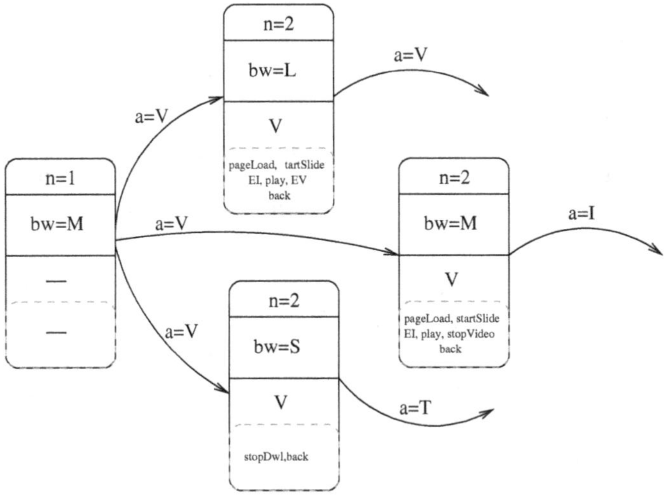

Fig. 12. Subset of optimal policy applied on MDP

Figure 12 provides an example of a part of the optimal policy obtained by *Q-learning*. Initially, for $bw = M$, the agent decides $a = V$ (video version). The stochastic dynamics of the context (i.e. user and resources) bring it in one of the possible next states. Three of them are illustrated in Figure 12.

The bottom state corresponds to a *bw* decrease (i.e. $M \rightarrow S$) and/or a smaller user interest inferred from the sequence of interactions *stopDwl,back*. The ensuing decision is therefore $a = T$. On the contrary, under better conditions (i.e. $M \rightarrow L$) and/or interested user (i.e. both images and video are fully consumed), the agent continues to propose the richest version (see the top state). Finally, for a medium interested user, the next decision is $a = I$.

These examples show that there exists a natural correlation between the wealth of the selected versions and the implicit user interest. For each movie, the choice of version is therefore based not only on available resources but also on the events observed while browsing previous movies. Hence, **user interactions sequence becomes an implicit descriptor of user interest**.

In general, the decision-making automaton depends on n and bw. When n, bw and It vary, the automaton becomes too complex to be displayed. The results of the POMDP require a different presentation. Henceforth working with **3 granularity levels on** n, **2 on** bw, **3 on** It and the set of rewards R_1 leads to a *policy graph* of more than 100 nodes. We apply it during numerous sequences of simulated navigations. Figure 13 gives the statistics on the decisions that have been taken. For every triplet (n,bw,It) the decisions - **the agent not knowing** It - are counted and translated into percentages.

We notice that the proposed content becomes statistically richer when the interest increases, proving again that the interest estimation from the previous observations is as expected. Let's take an example and consider the bottom-

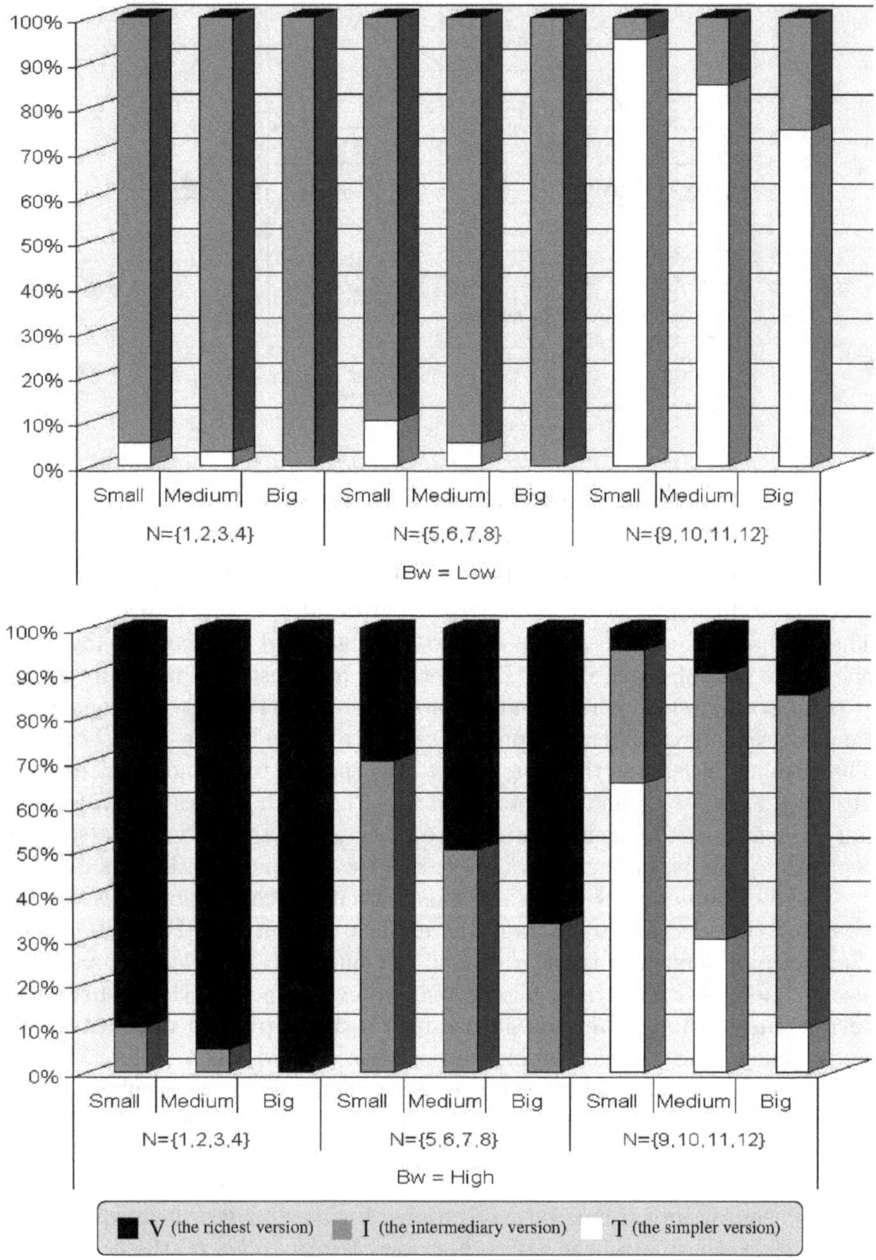

Fig. 13. Actions' distribution for the POMDP solution policy

right part of Figure 13 (corresponding to $BW=High$ and $N \in \{9, 10, 11, 12\}$). The probability of the policy proposing version V increases with the interest: from 4% (small interest) to 11% (average interest) then 14% (big interest).

Moreover, when n and/or bw increase, the interest trend is correct. For example, for a given set of It and n ($It=Medium$ and $N \in \{5, 6, 7, 8\}$), the proposed version becomes richer with the bandwidth's increase from (4%T, 96%I, 0%V) to (0%T, 51%I, 49%V). Additionally, from one set of rewards to another, these trends are always respected, although the values of the percentages are different.

7 Contextual metadata: contributions overview

Our contributions to the field of metadata-driven adaptation can be summarized as follows. Four conditions need to be fulfilled in order to benefit fully from our adaptation framework.

1. **Hidden metadata** We consider non-fully observable metadata (hidden contextual variables) that are useful to adapt a given multimedia content or service. Many semantic metadata (sometimes very subjective) are hidden: in this chapter, we have discussed user interest level and media importance descriptors (see Section 3.3).
2. **Optimizing adaptation decision-taking** We consider that adaptation decision-taking can be optimized sequentially. It is feasible to look for the best sequences of adaptation actions in order to improve quality of service or quality of experience. This implies the existence of a performance criteria (cumulated rewards) used to compare adaptation policies.
3. **Hidden metadata importance** Naturally, the more we know about hidden variables, the better the adaptation decision-taking should be.
4. **Inferred metadata** We also need a way to infer or sequentially estimate the values of hidden metadata from fully-observable contextual metadata. In our study, we use implicit feedback to infer hidden levels of user interest.

Under these four conditions, we have shown that our formal approach for reinforcement learning-based adaptation allows to automatically handle partially-observable semantic metadata. A decisional agent manages these metadata by memorizing fully-observable contextual metadata. Therefore the agent can be embedded in the content and can be considered metadata itself.

8 Conclusion

The chapter has presented an original agent-based adaptation approach. According to the state of the context, the agent decides sequentially what the best adaptation actions are. The MDPs/POMDPs form the theoretical background of the agent. The main contribution of this research work is the study of the impact of our approach on the contextual metadata. In particular we have shown a mechanism for manipulating partially observable descriptors.

These ideas have been applied to adapt a movie browsing service. In particular, we have proposed a method for refining an adaptation policy according to user interest. Our decision models naturally handle subjective metadata such as user interest. As a result, we are able to abstract user interest descriptors that are implicitly generated by observing user interactions.

The perspectives of this work are numerous. First, a more thorough integration of our approach with MPEG-21 should be studied. Second, extending the context state by taking other variables into account (e.g. semantics like genre, actors, etc.) should further optimize the decision making process. Third, compact, XML-like representation formats for adaptation policies should be proposed. Finally, since these policies are sequential and become heavier as the content complexity increases, it would be natural and useful to consider streaming them.

References

1. Mariam Kimiaei Asadi. *Multimedia Content Adaptation with MPEG-21*. PhD thesis, ENST Paris, June 2005.
2. Anthony R. Cassandra, Leslie Pack Kaelbling, and Michael L. Littman. Acting optimally in partially observable stochastic domains. In *Proceedings of the Twelfth National Conference on Artificial Intelligence*, volume 2, pages 1023–1028, 1994.
3. Vincent Charvillat and Romulus Grigoras. Reinforcement learning for dynamic multimedia adaptation. *Journal of Networking and Computer Applications*, 30(3):1034–1058, 2007.
4. Richard O. Duda, Peter E. Hart, and David G. Stork. *Pattern Classification (2nd Edition)*. Wiley-Interscience, 2000.
5. Dietmar Jannach, Klaus Leopold, Christian Timmerer, and Hermann Hellwagner. A knowledge-based framework for multimedia adaptation. *Applied Intelligence*, 24(2):109–125, 2006.
6. T. Joachims, L. Granka, and B. Pan. Accurately interpreting clickthrough data as implicit feedback. In *SIGIR'05*, August 2005.
7. D. Kelly and J. Teevan. Implicit feedback for inferring user preference : A bibliography. In *SIGIR Forum*, volume 37, pages 18–28, 2003.
8. Cedric Kiss. Composite capability/preference profiles (cc/pp): Structure and vocabularies 2.0, May 2007.
9. D. Mukherjee, E. Delfosse, Jae-Gon Kim, and Yong Wang. Optimal adaptation decision-taking for terminal and network quality-of-service. *IEEE Transactions on Multimedia*, 7(3):454–462, 2005.
10. Cezar Plesca, Vincent Charvillat, and Romulus Grigoras. Adapting content delivery to observable resources and semi-observable user interest. Research Report IRIT/RR–2007-3–FR, Toulouse, January 2007.
11. M.L. Puterman. *Markov decision processes: discrete stochastic dynamic programming*. Wiley-Interscience, 1994.

12. Pedro M. Ruiz, Juan Botia, and Antonio Gomez-Skarmeta. Providing qos through machine learning driven adaptive multimedia applications. *IEEE Transactions on Systems, Man and Cybernetics*, 34(3):1398–1411, 2004.

13. Guy Shani, David Heckerman, and Ronen I. Brafman. An mdp-based recommender system. *Journal of Machine Learning Research*, 6:1265–1295, 2005.

14. Satinder P. Singh, Tommi Jaakkola, and Michael I. Jordan. Learning without state-estimation in partially observable markovian decision processes. In *International Conference on Machine Learning*, pages 284–292, 1994.

15. R.S. Sutton and A.G. Barto. *Reinforcement Learning: An Introduction*. MIT Press, 1998.

16. T. Syeda-Mahmood. Learning and tracking browsing behavior of users using hidden markov models. In *IBM Make It Easy Conference*, 2001.

17. Anthony Vetro, Christian Timmerer, and Sylvain Devillers. *Digital Item Adaptation - Tools for Universal Multimedia Access*. 2006.

Part III

Multimedia Retrieval

Semantics in Content-based Multimedia Retrieval

Horst Eidenberger[1], Maia Zaharieva[2]

Vienna University of Technology, Vienna, Austria
[1] `eidenberger@tuwien.ac.at`
[2] `maia@prip.tuwien.ac.at`

This contribution investigates the content-based feature extraction methods used in visual information retrieval, focusing on concepts that are employed for the semantic representation of media content. The background part describes the building blocks of feature extraction functions. Since numerous methods have been proposed we concentrate on the meta-concepts. The building blocks lead to a discussion of starting points for semantic enrichment of low-level features. The second part reviews features from the perspective of data quality. A case study on content-based MPEG-7 features illustrates the relativity of terms like "low-level," "high-level" and "semantics". For example, often more semantics mean just more redundancy. The final part sketches the application of features in retrieval scenarios. The results of a case study suggest that – from the retrieval perspective, too – "semantic enrichment of low-level features" is a partially questionable concept. The performance of classification-based retrieval, it seems, does hardly depend on the context of features.

1 Introduction

This contribution discusses the role of semantics in content-based image retrieval [12, 17]. It is organised in three sections. In the first section, we focus on *what (semantic) content-based features are*. Basic building blocks of signal processing-based low-level features are reviewed and starting points for enrichment of features are discussed and experimentally evaluated. The second section describes *what* features extract. It sketches analysis methods that allow for looking behind the scenes of feature extraction. Technically, the quality of feature extraction methods is judged from a quantitative point of view. Based on these insights, the third section

H. Eidenberger and Maia Zaharieva: *Semantics in Content-based Multimedia Retrieval*, Studies in Computational Intelligence (SCI) **101**, 151–174 (2008)
www.springerlink.com

illustrates *how features are used*. It reviews application scenarios and – in a case study – compares the performance of content-based low-level features to context-free features in a typical retrieval scenario.

The paramount intention of this contribution is to give the reader a feeling for *how* content-based image features work and what enrichment with semantic information *means* (especially, on the data level). It is shown that, in practice, the borders between low-level feature extraction, semantic and context-free features almost vanish. For example, quantisation methods – frequently used in most image features – are nothing else but the introduction of domain knowledge into the feature extraction process. In this context, "domain" denotes concepts on a number of levels. It may stand for application domains (regular application case for semantic methods) but as well for technical domains (e.g. image properties). The essence of successful semantic image retrieval is to learn to handle the trade-off between such constraints and the benefit from narrowing the ambiguity of media content. Eventually, image retrieval is an ill-posed problem. For every gain in performance a considerable amount of generality has to be given away.

2 Building blocks of content-based image features

2.1 Signal Processing Building Blocks

In recent years, a remarkable number of content-based image features has been proposed, ranging from simple description of colour distributions to description of complex objects [12]. The visual part of the MPEG-7 standard [16] provides a comprehensive overview over the state of the art in low-level feature extraction. The building blocks of content-based features can be split into five groups (see Figure Fig. **1**). Each building block provides enriched information that is further processed in consecutive data manipulation steps.

1. *Unitary time-to-frequency transformations* create a frequency representation of the media signal. Frequencies describe fundamental variations in the visual content. Low frequencies correspond to smooth changes (e.g. large uniform areas of background) and high frequencies correspond to abrupt changes (e.g. edge and corner information, but also noise). The Discrete Fourier Transformation (DFT) is the classic representative of such transformations. It is widely employed in audio retrieval. In visual retrieval, the Discrete Cosine Transformation (DCT,

e.g. compression) [19], Wavelet-based multi-resolution analysis [14] and the Angular Radial Transformation (ART) are more common. In the MPEG-7 standard, for example, DCT and ART are adopted in *Color Layout* descriptor and the *Region-Based Shape* descriptor, respectively [8].

2. *Parametric image transformations* map the input data to a condensed space. The mapping is controlled by parameters and usually not invertible. One example for parametric transformations is the Hough transformation [7]. It is employed to detect lines and to extract objects. The Radon transformation [21] is a second example for a parametric transformation. It is used to provide a rotation-invariant representation of image content. For this purpose, it is, for example, employed in MPEG-7 in the Homogeneous Texture descriptor that extracts characteristics of textures.

3. *Localisation methods* are used to select and describe data regions for further processing. State of the art techniques include the segmentation into rectangular macro-blocks, the definition of geometric primitives as regions of interest, spatial segmentation methods (e.g. by boundary detection, edge operators, etc.) and frequency filtering. The selection of the appropriate technique is mostly determined by the specific characteristics of the input data and the consecutive processing steps. MPEG-7 uses several localisation descriptors such as the Region Locator for regions of interest or the Contour-based Shape descriptor for spatial segmentation.

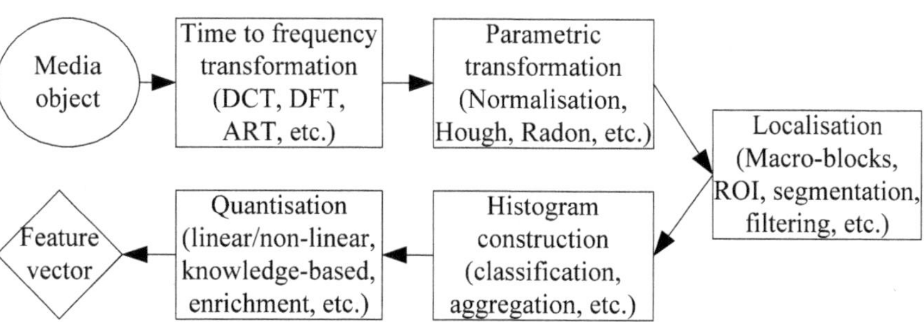

Fig. 1. Royal order of feature extraction building blocks.

4. *Histogram methods* are the methods of choice for data unification and conceptual understanding of types and frequencies of occurrence. They are conceptually similar to parametric transformations (e.g. they are not invertible). Based on (semantic) similarity criteria, data is

grouped (clustered) and counted. Such aggregation methods are not only applicable for colour properties and intensity distributions, but for any data population that contains a minimum of uniformity (e.g. edge types, frequencies, etc.). One application example in MPEG-7 is the Edge Histogram descriptor that provides information about the distribution of edge types in localised image regions.

5. *Quantisation methods* reduce and/or normalise the amount of high-frequency information in the feature data. Like aggregation, quantisation requires a certain amount of semantic knowledge. Quantisation is, for example, used for the reduction of storage space and for the satisfaction of limited bandwidth requirements. Methods include simple linear quantisation (e.g. normalisation of colour ranges by division into equally sized intervals), non-linear quantisation (e.g. variable intervals), and various types of "magic" quantisation. *Magic quantisation* puts an umbrella over all methods that consider additional (semantic) information such as domain rules, user information, etc. The bin-wise quantisation rules employed in the MPEG-7 *Scalable Color* descriptor are good examples for magic quantisation.

2.2 Starting Points for Semantic Enrichment

One of the most relevant ongoing activities in feature design is the semantic enrichment and interpretation of low-level features. Semantic enrichment endeavours to narrow the semantic gap. Since the perception of visual content is highly subjective, the same content may be interpreted differently by different users or even by the same user under different circumstances. Generally, human perception is based on three types of stimuli: generally perceived (not recognised) stimuli (e.g. colour or intensity distribution), specifically perceived (recognised) stimuli (e.g. object recognition), and pseudo-random stimuli (e.g. psychological, sociological, etc.). Retrieval of media objects exclusively by generally perceived properties is unsatisfactory. Mainly three sources of information are available for feature enhancement: information on the application domain (e.g. on media content), information on the user (e.g. user or retrieval preferences), and information on the characteristics of the specific features (e.g. statistical properties). Semantic feature research comprises the following (selected) groups of approaches:

1. *Feature data enrichment.* Additional knowledge can be induced (mainly in the aggregation and quantisation steps) by methods from statistics, artificial intelligence, neuronal networks, etc. For example,

domain knowledge on football could be used to identify field, ball and players from shape features (e.g. circularity). Application knowledge on the arts could be used to distinguish between classic and surreal paintings based on colour and edge features. Currently, many research efforts focus on the reduction of the search space by clustering [9]. Most clustering methods are unsupervised classification techniques (e.g. hierarchical (agglomerative or divisive) clustering [9], Self-Organizing Maps [11]). Such approaches divide the set of processed data (image features) into subsets (clusters) based on inherent data similarities (measured as metric distance). Hence, each cluster contains objects of more than average similarity.

2. *Context-based querying.* Integration of the media context in the feature extraction process is another starting point of semantic retrieval. Often, images are embedded in some type of document (e.g. web pages, multimedia presentations, text documents). In such situations, multimodal feature extraction, enrichment and retrieval based on semantic concepts may improve the relevance of media descriptions significantly. The joint application of text retrieval, content-based visual and audio retrieval can be performed by clustering, co-quantisation and during query execution.

3. *Feature knowledge modeling.* Eventually, though today semantic web technologies are almost exclusively employed for the annotation of markup text and for knowledge representation (e.g. RDF Schema, OWL, DAML+OIL), in the future they may also find applications in semantic enrichment of content-based media descriptions. Ontologies offer meta-concepts for the description of constraints and relationships among objects (for example, cardinality, domain and range restrictions, union/disjunction, inverse rules, etc.). In the context of content-based features ontology languages could be applied to specify constraints and relationships among descriptions, feature elements and semantic knowledge. Content-based ontology concepts would allow a clearer distinction between feature extraction models, descriptions and semantic metadata. Furthermore, they could provide the basis for rule-based feature derivation (e.g. localisation relationships, quantised colour histograms). Currently, such models are barely available. Many content-based features are a mixture of signal processing and the application (often, without reflection) of semantic knowledge.

3 Feature structure analysis

The first section has introduced how features are extracted from the media content and – in principle – how features can be lifted to a semantically higher level. This section investigates features from the perspective of data quality. Statistical data analysis offers tools that deserve more attention in visual information retrieval since these methods allow, for example, to judge whether an enrichment method causes more than just a higher level of information redundancy.

3.1 General-Purpose Feature Analysis Methods

Below, it is explained how quantitative data analysis methods can be employed to evaluate content-based media. It is shown how statistical analysis methods can be applied to judge various aspects of feature data quality. For example, cluster analysis methods and factor analysis methods (e.g. Principal Component Analysis) can be used to identify redundancies in the feature data. Topological clustering techniques are valuable tools to test the sensitivity of features for changes in the media data (e.g. noise, data loss). Statistical indicators can be applied to express the quality of feature data distribution and hence, the potential of features for similarity-based discrimination in retrieval.

The traditional evaluation scheme used in (visual) information retrieval employs recall and precision indicators computed for sample queries on well-known media collections. This process has its drawbacks and leaves the feature designer with a handful of open problems. The most relevant issue for application is defining a *ground truth* that reflects human similarity judgement appropriately independently of cultural aspects and other human peculiarities. Analysis of a feature requires embedding it in a querying framework and execution of hundreds of queries in order to guarantee statistical validity of the quality indicators. This process has to be repeated on every change in the feature transformation and it does not provide any hints on problems in the feature transformation. The bottom line is that recall and precision allow to estimate whether a feature performs well but not *why* it (which of its (semantic) properties) shows the observed behaviour.

9.3.1.1 The Big Picture of Quantitative Feature Analysis

The proposed supplementary evaluation procedure is a lightweight process that employs statistical data analysis methods (especially, Self-Organizing

Maps [11]) to evaluate information-theoretic quality aspects of feature transformations. A querying framework and ground truth information are not required. Quality indicators are derived directly from the feature data and allow conclusions on the statistical quality of the extraction process.

As for the traditional approach feature transformations are applied on predefined media collection. The resulting *feature data matrix* (feature vector elements by media objects) is normalised and investigated for characteristic properties (e.g. variance per variable), common properties and similarities by statistical methods. The essential point for understanding the key idea is that – in contrast to the retrieval situation – here, media objects are used to describe feature elements (e.g. colour histogram bins, shape moments). That is, technically, statistical methods are applied on the *transposed* feature data matrix. In statistical terms, the feature vector elements are the variables and the media objects are the cases. For example, in the experiments below MPEG-7 description elements (*Color Layout* coefficients, *Edge Histogram* bins, etc.) are fed – all described by the same set of media objects – into a Self-Organizing Map clustering algorithm to identify similarities between them.

Apart from applying statistical methods on visual features, another new aspect of the proposed evaluation process is that the feature in question is not just compared to itself but also to *reference data*. This reference data is provided by the content-based visual part of the MPEG-7 standard. Visual MPEG-7 descriptions are calculated for the predefined media collections. Comparing the feature vectors of the evaluated feature transformation to the reference data allows for gaining additional insights on the characteristics of a feature.

This proposed procedure has several advantages: Firstly, measurement is performed in a systematic way: one system (feature) is compared to another. Since the process is independent of the user (no user input required) it yields objective results. Secondly, results are application-independent. General data quality is measured instead of retrieval quality. Furthermore, no querying framework is required to apply this evaluation method. All necessary steps can be fulfilled with mathematical/statistical standard software (e.g. SPSS, Matlab).

3.1.2 Application Scenarios

A variety of questions including the following can be answered by statistical analysis:

- What is the "type" of a new feature? With respect to the MPEG-7 visual norm, is it a colour, texture, shape or motion feature or does it define en-

tirely new (semantic) criterions for visual media? Investigating prox-
imities between a new feature and visual MPEG-7 descriptions may
give valuable indications on promising starting points for closer (e.g. al-
gorithm-based) examination of the new feature and help avoiding un-
wanted parallel developments.

- How robust is a new feature against rotation, scaling and other visual
  media transformations? Are the feature vector elements of original and
  transformed media data still similar after transformation? If not, do
  transformations change the characteristics of the feature transform?
- How robust is a new feature against noise? Do the characteristics of the
  feature vectors change if the media objects are noisy? This includes coding noise, i.e. artefacts introduced by lossy coding algorithms. Of
  course, sensitivity for every other type of noise can equally be tested, if
  the required media data is available.
- What is the effect of semantic enrichment on the data quality? Does an
  enriched feature represent new properties that are independent of those
  already identified by low-level features or does the semantic enrichment
  just cause higher data redundancy or make the feature extraction procedure more noise-prone?
- Does a feature mapping represent human visual similarity perception
  adequately? If the feature transformation is applied to two collections of
  similar media objects, are the corresponding feature vector elements
  similar, too?

Apparently, stronger answers can be given on these questions, if – in addition to recall- and precision-like evaluation procedures – statistical methods are considered.

3.1.3 Evaluation Workflow, Data Basis and Reference Data

The flow of work in the statistical evaluation procedure is as follows.
Firstly, the new feature transformation is applied to predefined media collections. Numerical feature vectors are extracted. In the second step these
feature vectors are merged with the pre-extracted MPEG-7 descriptions.
After merging, data are normalised to a certain interval (e.g. [0, 1]) or particular moments (e.g. $\mu=0$, $\sigma=1$). Conventionally, all feature vectors together are addressed as the *feature matrix* (feature elements in columns,
media objects in rows). On the feature matrix statistical operations are applied and indicators are derived. In the last step, these indicators are visualised and interpreted. Based on the interpretation the proposed feature transformation can be iteratively refined.

Various statistical methods exist that can be employed for evaluation. Principally, the three main areas relevant for visual information retrieval are univariate description, detection of similarities and detection of dependencies in the feature matrix (both multivariate). In earlier experiments we found these methods very useful [5]:

- Extraction of moments of first and second order of feature elements as well as computation of a discrete distribution of values for each element. The distribution reveals how often each value (down-sampled to a few bits) occurs. For example, it allows conclusions on how well a feature element utilises its data type.

- One- and two-dimensional cluster analysis of *feature elements* (not media objects!) for similarity assessment. *K*-means clustering and dendrograms for visualisation have proven to be sufficient in the one-dimensional case. Unfortunately, dendrograms become soon unreadable for larger numbers of elements. In this situation, two-dimensional clustering techniques (e.g. Self-Organizing Maps [11]) yield better results.

- Detection of dependencies of feature vector elements by factor analysis. Eigenvalues extracted from a data matrix (e.g. by a Principal Component Analysis) can be interpreted as hidden factors that have a linear influence on the data values. Elements (media properties) that are significantly influenced by the same factors (expressed by a factor loadings matrix) are obviously dependent of each other.

Principally, any media collection can be used for statistical analysis. The definition of a ground truth is not required. Still, it simplifies the interpretation process (especially, if semantic features are concerned), if the media collections have an inherent context. For some statistical evaluation methods only testbeds with a small number of media objects are manageable. Furthermore, statistical evaluation results depend to a certain degree on the a priori structure of the investigated media collection. Even though the media basis may be chosen arbitrarily, its structure has to be taken into account as a biasing factor in the interpretation of results (e.g. if cluster analysis is used)!

3.1.4 Summary

Semantically enriched features should have significantly different statistical properties than low-level features. Semantic enrichment desires to induce more meaning into the feature data. However, applying the proposed statistical analysis techniques reveals that often, this means just more redundancy. In simple words, semantically enriched features often look the same for different media content. The negative consequences of this effect for retrieval are obvious.

3.2 Case Study: Structure Evaluation of MPEG-7 Features

A few examples should make the suggested analysis approach and its advantages more transparent. Below, the statistical method introduced in the previous subsection are employed to judge the (semantic) quality of feature data extracted by selected content-based visual MPEG-7 descriptors. The analysis focuses on redundancies (e.g. from the application of semantic quantisation methods) and on sensitivity to content changes and noise.

3.2.1 Case Study Setup

We analyse the majority of the content-based visual MPEG-7 descriptors. All colour descriptors: *Color Layout, Color Structure, Dominant Color, Scalable Color*, all texture descriptors: *Edge Histogram, Homogeneous Texture, Texture Browsing* and one shape descriptor: *Region-based Shape*. The other basic shape descriptor, *Contour-based Shape*, is not used, because it produces structurally different descriptions that cannot be transformed to data vectors measuring on interval scale. Description extraction is performed employing the MPEG-7 experimentation model (XM, [23]) of MPEG-7 Part 6: Reference Software. In the extraction process each descriptor is applied on the entire content of every media object.

The descriptors are applied on three image collections: the Brodatz dataset [2] (112 monochrome images, 512x512 pixel), a subset of the Corel dataset [24] (260 colour photos, 460x300 pixel, portrait and landscape) and a dataset with coats-of-arms images [1] (426 synthetic images, 200x200 pixel). The evaluation is performed in the following steps: description extraction, normalisation, extraction of statistical indicators, quantisation and extraction of distributions, hierarchical cluster analysis, computation of topological cluster structures and factor analysis. After the description extraction, the resulting XML-descriptions are transformed into a data matrix with 798 lines (media objects) and 314 columns (description elements).

Mean and standard deviation are used as primary indicators for description elements. To identify the distribution of values of description elements over N media samples, the coefficients of the data matrix are quantised to ten bins. For the hierarchical cluster analysis a single-linkage algorithm with squared Euclidean distance measurement is used. The results are depicted as dendrograms on a relative scale from 0 (identical) to 25 (not similar). Self-Organizing Maps (SOMs) [11] are employed for topological clustering. SOMs are calculated with a hexagonal layout (every non-border cluster has six neighbours). For cluster adaptation, a Gaussian neighbourhood kernel is employed. Maps are initialised randomly. For factor extraction a Principal Component Analysis (analysis of the coefficients of the correlation matrix) is used [13]. All Eigenvalues greater than one are selected as factors. To simplify the interpretation process, a Varimax rotation is performed on the factor loadings matrix. Factor analysis can only be applied on elements with existing variance. For the Brodatz dataset 225 description elements fulfil this requirement, for the Corel dataset 311 and for the coats-of-arms dataset 310. For the remaining elements, the description extraction algorithms comes up with exactly the same values independent of the analysed content.

3.2.2 Redundancy Analysis

In this analysis we try to identify whether the description elements extracted from visual content are *unique* or not. Redundancy information is highly valuable for two major reasons. It may influence how descriptions are organised in description schemes (*efficiency of application*). It is obviously not desired to combine certain descriptors to a description scheme (e.g. as a means of semantic enrichment) if it is well known that the descriptors are highly redundant for the concerned media class. Additionally, it can be used as a supplementary method to the MPEG-7 binary format [16] for compression of descriptions (e.g. for specific classes of content). This helps to further reduce the amount of space and bandwidth needed in visual retrieval systems (*efficiency of representation*).

A first striking result revealed by the hierarchical cluster analysis is the high self-similarity of the elements of the *Homogeneous Texture* descriptor for any type of media (see Table 1). For the Brodatz dataset (rich textures) and the coats-of-arms dataset (poor textures) all description elements form a single cluster with a maximum distance of 4%. Interestingly, the *Edge Histogram* descriptor forms five to ten clusters with ten to 15 elements for any type of content. The elements of these clusters are self-similar but the distance between the clusters is relatively large.

Table 1. Results of hierarchical cluster analysis: number of clusters and distances between clusters. The maximum distance is given in percent (where 100% would be the distance of a vector of "0" values to a vector of "1" values).

Descriptor	Media collection	No. of clusters	Maximum distance between clusters
Homogeneous Texture	Brodatz, Coats-of-arms	1	4%
	Corel	2	20%
Edge Histogram	any	5-10	12%-20%
other	any	>5	>20%

Fig. 2. Self-Organizing Map of MPEG-7 description elements for the coats-of-arms dataset. Neighbouring clusters contain similar description elements. Since every non-border cluster has six neighbours, clusters are shown as hexagons. Cluster populations are depicted as textures (CLD: *Color Layout*, CSD: *Color Structure*, DCD: *Dominant Color*, EHD: *Edge Histogram*, HTD: *Homogeneous Texture*, RSD: *Region Shape*, SCD: *Scalable Color*, TBD: *Texture Browsing*). If clusters are shared between descriptors, hexagons are split into triangular regions.

Analysing the SOMs for the three media collections (see Figure Fig. **2**) supports the first impression of the hierarchical analysis. *Homogeneous Texture* lays a fine-meshed net over the investigated media property. The *Edge Histogram* descriptor forms clusters that contain slightly more elements. The net of the *Edge Histogram* is wide-meshed but the descriptor covers a larger area of the variance in the media data. All other descriptors form rather small 2d clusters for any type of content.

A more detailed view can be obtained from the factors extracted by factor analysis algorithms (see Table 2). For the Brodatz dataset 34 factors explain 225 description elements (the remaining elements have zero variance). It is surprising that the MPEG-7 descriptors perform slightly worse on the coats-of-arms dataset than on the Corel dataset. The Corel photos contain more details and, generally, descriptions should be less redundant on material with richer content. Another interesting result of the factor analysis is that – for any type of content – the *Dominant Color* descriptor has the tendency to identify colours with identical colour component values. Maybe certain characteristics (e.g. quantisation) in the extraction algorithm implemented in the MPEG-7 XM cause this phenomenon.

Table 2. Results of factor analysis: number of extracted factors and explained variance. Only elements with existing variance are considered.

Media collection	Elements with existing variance	Factors	Explained variance (all)	Explained variance (first factor)	Redundancy relationship
Brodatz	225	34	89%	15%	7:1
Corel	311	69	85%	12%	9:2
Coats-of-arms	310	71	80%	6.7%	9:2

Several observations can be made from these analysis results. Generally, the MPEG-7 descriptors generate results of high redundancy. The *magic quantisation by feature characteristics* used in most descriptors may be one explanation for this observation. Especially, all MPEG-7 descriptors are highly redundant for monochrome media content. On the other hand, all bins of *Color Layout* are highly un-similar for any type of media content and independent from all other elements. Especially the first element (luminance DC coefficient) seems to be a good indicator for global shape information even for complex scenes (as *Region-based Shape* should be). The elements of the *Homogeneous Texture* descriptor are – independent of the media – highly self-similar and redundant. The ideal – content-independent – description scheme for visual content seems to be *Color Layout* (because of the first element), *Dominant Color*, *Edge Histogram* and *Texture Browsing*. This DS provides a maximum of semantic context (on the MPEG-7 level) at a minimum of redundancy.

3.2.3 Sensitivity Analysis

This analysis tries to give indication on the sensitivity of the descriptors on varying media content. In detail, three forms of sensitivity are investigated: firstly, sensitivity of colour descriptors for monochrome content, secondly, sensitivity of colour descriptors for content with few colour shades (e.g. animations) and finally, sensitivity of the texture descriptors and *Region-based Shape* for coarse, medium and fine structures in the content. Ideally, the descriptors should provide surjective mappings from the visual content to the feature space. These mappings should be robust against variations in the quality of the content (e.g. presence of colour information, resolution). Analysing the sensitivity allows to judge to which extent "bad" (e.g. bleached) input affects the quality of the descriptions.

The main indicators for sensitivity are mean and standard deviation. For a uniformly distributed element on the interval [0, 1] with a mean of 0.5, the maximum standard deviation is 0.346. In the evaluation the standard deviation should be 0.2 or higher (using at least an interval of 40% of the data range for 66% of all media objects) to be acceptable. Then, the description element can be considered as being sufficiently discriminant to distinguish media objects independently of variations in the content.

Table 3 summarises the average means and standard deviations of the colour description elements. *Color Layout* performs badly on monochrome data (Brodatz dataset). Only six of twelve bins have a standard deviation greater than zero: the DC and AC coefficients of the luminance channel. Whenever colour is present – independent of the number of gradations – *Color Structure* performs excellently. The average standard deviation is 0.25. Therefore, the element values are distributed over the entire range of possible values. For monochrome content, *Scalable Color* is not able to derive meaningful descriptions. For the Corel dataset, *Scalable Color* results are excellent. The average standard deviation is 0.3. *Edge Histogram* performs excellently on any type of media (see Table 4 for details on texture and shape descriptors). The *Homogeneous Texture* descriptor performs poorly on colour images, especially if they have few colour shades and textures in them. Finally, the *Region-based Shape* descriptor measures excellently on any type of media. These findings are supported by the cluster analysis results. Most clusters are on distance level lower than 20%. Hardly any clusters exist at average distance (20% to 60%). Cluster structure and clusters size varies widely for different content.

Table 3. Average mean and standard deviation of colour description elements. Only elements with existing variance are considered in the averaging process.

Descriptor	Media collection	Average mean	Average standard deviation
1. *Color Layout*	Brodatz	0.7	0.1
	Corel	0.55	0.2
	Coats-of-arms	0.65	0.15-0.2
Color Structure	Brodatz	0.85-0.9	0.05-0.15
	Corel, Coats-of-arms	0.5	0.25
Dominant Color	any	0.45-0.5	0.3
Scalable Color	Brodatz	0.4	0.3
	Corel	0.5	0.3
	Coats-of-arms	0.4-0.5	0.15

Table 4. Average mean and standard deviation of texture and shape description elements.

Descriptor	Media collection	Average mean	Average standard deviation
Edge Histogram	any	0.5	0.25-0.3
Homogeneous Texture	Brodatz	0.65-0.7	0.1
	Corel	0.75	0.1
	Coats-of-arms	0.75	0.05
Texture Browsing	Brodatz, Corel	0.2-0.3	0.2-0.25
	Coats-of-arms	0.1	0.05
Region-based Shape	any	0.5	0.2-0.25

3.2.4 Summary

All colour descriptors work excellently on photos (high-frequency input) but *Color Layout, Color Structure* and *Scalable Color* perform badly on artificial media objects with few colour gradations, and very badly on monochrome content. *Edge Histogram* is by far the best texture descriptor (high sensitivity, low redundancy). *Homogeneous Texture* is highly sensitive to the analysed media content and the variance of results is small. *Region-based Shape* is a good descriptor that can be applied to any type of media content.

These results are partially surprising. MPEG-7 descriptions are highly redundant, sensitive to noise and provide only little ground for discrimination. Especially, the quantisation methods proposed in MPEG-7 (semantic enrichment based on feature knowledge) cause a regrettable loss in the quality of media descriptions. This behaviour is a striking example for the dangers of semantic enrichment. Descriptions may become less discriminant and drift to represent superficial media properties. One possible cure for such structural problems would be to revise the application of additional knowledge in concerned descriptors (e.g. the quantisation steps suggested in the MPEG-7 standard).

4 Features at work

4.1 (Semantic) Feature-Based Retrieval Approaches

Multimedia information retrieval systems are often distinguished by the querying paradigm. Frequently used methods are querying by example, querying by sketch, iconic querying, and querying by example groups. Though being important for the user, the selected querying paradigm does not influence the retrieval process. Media comparison is always based on descriptions (feature vectors). If not available beforehand, descriptions have to be extracted during the querying process (e.g. of sketches or example images).

In most systems, one of two retrieval processes is employed: retrieval based on the *vector space model*, or retrieval based on *probabilistic inference*. The vector space model assumes that media descriptions are points in a vector space and that this vector space has a geometry (mostly, Euclidean geometry is assumed) [6]. Then, unsimilarity of media objects can be measured as metric distance of media descriptions (e.g. City Block distance, Euclidian distance, Minkowski distances). The vector space model is successfully used in text information retrieval. Unfortunately, applied to MMIR, two problems arise. Firstly, it is not clear what type of geometry (distance function) fits to human similarity perception. Secondly, often differently extracted media descriptions require different distance measures. The selection of features for retrieval and the usage of multiple distance measures are non-trivial, still open research problems (see, for example, [4]).

Probabilistic inference models use media descriptions and a priori probabilities (computed from statistics based on e.g. human relevance information) to compute differentiated a posteriori probabilities that can be used for retrieval [6]. Employed models are mostly based on Bayesian networks, i.e. topologies that represent dependencies among features. The major advantage of probabilistic inference models over the vector space model is that they avoid the problem of explicitly defining similarity measures. The main disadvantages are that sample data are required and that fast-learning relevance feedback algorithms (see below) are hard to define. Furthermore, the *independence constraint* assumed in many inference models (e.g. the Binary Independence Retrieval Model [6]) may be too restrictive for real-world scenarios. Human perception is highly sensitive and culturally loaded. For example, a picture of two adults with a child may be perceived differently from the same picture without the child (family vs. lovers).

Severe problems as the semantic gap and polysemy have lead to the insight that visual retrieval may be modelled best as an iterative process. Retrieval steps should be directed by the user's relevance feedback. Today, one refinement technology outperforms most other approaches: classification by Support Vector Machines (SVM) [18]. SVMs separate a given set of binary labelled data (relevant/irrelevant) by their maximum margin, i.e. a hyper-plane in maximum distance of the two groups. Moreover, kernel functions are employed to project the input data to a higher-dimensional (less densely populated) feature space. This transformation simplifies the separation process. The advantages of SVMs are straightforward application and high performance. SVMs are easy to apply, since only two groups (relevant and irrelevant media objects) have to be distinguished by the user. The outstanding performance of SVMs may at the same time be their major weakness. SVMs are prone to overfitting. Hence, careful selection of training data is a crucial step in SVM application.

It should be noted that none of these retrieval approaches makes assumptions on the *semantic level* of the employed media metadata. As long as certain mathematical requirements are fulfilled (e.g. measurement on interval scales), the feature data may be of arbitrary shape. In experimental evaluation it may turn out that semantic concepts and data quality considerations are eventually irrelevant for retrieval application. This suspicion may especially be true in cyclic retrieval scenarios based on classification. The following case study evaluates this question.

4.2 Case Study: CB Features vs. Semantic Features

In this section we turn the considerations from the previous sections to the extreme and regard context-free (random) features as semantic features, i.e. as expressions of visual content that are human-like, hence (almost) unpredictable for machines. In a retrieval scenario, we compare the performance of the context-free image features to sophisticated signal processing-based features (the MPEG-7 image features [16]). Our context-free features show a maximum of the data anomalies investigated in the evaluation section. However, they are employed in the same retrieval processes as the MPEG-7 descriptions.

4.2.1 Case Study Setup

The experiments are undertaken in an environment that simulates user-centred visual information retrieval. Querying is performed as cyclic refinement of results by relevance feedback. That is, retrieval results are provided by classification. Kernel-based Support Vector Machines [25] are employed for active learning of classifiers. In the automated evaluation process the training samples are taken from a pre-defined ground truth (representing the visual similarity judgement of users). For the sake of realistic results only few examples are used for training. The trained classifiers are employed to discriminate the populations of queried collections into groups of relevant and irrelevant media objects.

The extraction of MPEG-7 features is performed by using the MPEG-7 experimentation model. The context-free features are extracted from the test dataset using the Matlab algorithms discussed in the results section. SVM-Light [10] is employed for SVM classification as well as for the computation of recall and precision values.

Training is performed on the fraction of the data (random selection of vectors) given in the results section (1%, 10% etc.). Classification is always performed on the entire ground truth group of positive examples and an equal number of negative examples. This rule allows for easier interpretation of results, since every feature below 50% recall/precision can be discarded for being dominated by guessing.

The following content-based visual MPEG-7 descriptors are employed in the analysis process [16]: *Color Layout, Color Structure, Dominant Color, Edge Histogram, Homogeneous Texture, Region-based Shape* and *Scalable Color*. In total, every media object is described by 306 feature elements. Two collections are merged to form the test media dataset: parts of the well-known Corel dataset (1615 colour images) and the UCID dataset provided by the University of Nottingham [22] (1338 colour images). The total 2953 images are split into *17 groups of conceptually similar content*. Noteworthy, the largest ground truth groups contain eight times more images than the smallest groups.

4.2.2 Experimental Results and Interpretation

Algorithm 1 computes a simple context-free feature for the media objects in the test dataset. 300 uniformly distributed random numbers per media object are employed as features. The dimensionality has been chosen to be comparable in length to the MPEG-7 features. Table 5 (1% training data) lists the performance of the semantic/random feature in relation to MPEG-7.

```
function extractSemanticFeature
```

```
inHandle = fopen('mediaFiles.dat');
data=zeros(1:1);
i=1;
while(feof(inHandle)==0)
    fname = fgetl(inHandle)
    for j=1:300
        data(i, j) = rand;
    end
    i=i+1;
end
fclose(inHandle);
```

Alg. 1. Extraction of simple random features.

Table 5. Performance of Alg. 1 in comparison to MPEG-7 (1% training data).

Feature	Kernel	Recall		Precision	
		%	Rank	%	Rank
MPEG-7	Linear	77.55	2	57.91	3
Random (A1)	Poly	74.61	4	51.02	9
MPEG-7	Poly	74.41	5	59.58	1
MPEG-7	Radial	73.76	6	59.21	2
Random (A1)	Linear	64.82	8	51.20	8
Random (A1)	Radial	62.44	9	51.52	7

```
function extractSemanticFeature
    % […] see Algorithm 1
    for j=1:300
        data(i, j) = i/numberOfFiles*rand;
    end
    % […] see Algorithm 1
```

Alg. 2. Extraction of improved random features.

The random feature performs inferior to MPEG-7 if sufficient training data is available. For 10% training data (data table not given here), recall is 11% and precision is 17% behind. Still, in situations where just a handful of training samples are available, the gap to MPEG-7 shrinks frighteningly. For the best kernel and 1% training data the random feature is only 3% recall behind the MPEG-7 features. These findings suggest that in SVM-based content-based features (semantic or not) are just a ground for discrimination. However, the more training samples are available, the larger the gap to the top performing features becomes. In terms of precision the random feature is always very close to the guessing border (50%). It seems that the random feature does not allow for efficient discrimination.

Our suspicion is that this deficit is caused by the uniform distribution of random numbers used (high redundancy!). In average, all random feature vectors are relatively similar.

Hence, in a verification experiment an improved random feature is employed to compete with MPEG-7. In equation 1, the feature f_i of the ith media object is a vector of j columns. N is the number of media objects in the queried collection. $random()$ returns a uniformly distributed random number. This feature associates media objects with random numbers of significantly different variances (frequencies). Alg. 2 shows the Matlab formulation of the feature extraction algorithm. Since Eq. 1 combines the random function with a step function, the feature is called $random\ step$ in the results in Table 6.

$$f_i = \left\langle \frac{i}{N} random() \right\rangle_j \tag{1}$$

The averaged performance indicators show that the random step feature performs astonishingly well. For 1% training data it outperforms MPEG-7 in terms of recall by 10%. Even its precision is competitive with just 1.5% behind MPEG-7. If 10% feature vectors are available for training (data table not given here), the difference in recall is still 9% and precision is head to head to MPEG-7. Astonishingly, random step matches the precision of MPEG-7. Since it is superior in terms of recall, the surprising conclusion of our experiment is that this simple "semantic" feature fulfils its purpose just as content-based MPEG-7 descriptions do.

Table 6. Performance of Alg. 2 in comparison to MPEG-7 (1% training data).

Feature	Kernel	Recall		Precision	
		%	Rank	%	Rank
Random Step (A2)	Poly	87.40	2	58.09	5
MPEG-7	Linear	77.55	3	57.91	6
Random Step (A2)	Linear	77.44	4	66.42	1
Random Step (A2)	Radial	76.80	5	61.70	2
MPEG-7	Poly	74.41	7	59.58	3
MPEG-7	Radial	73.76	8	59.21	4

From what we have learned in these experiments we advocate to shift the feature-related expectations of user-centred content-based retrieval. The role of media features is mainly to provide a *ground for discrimination* of user-labelled (similar and unsimilar) media objects. It is only of minor relevance if this ground is to 100% derived from media content. (This is a silent assumption in semantic enrichment, anyway.) Therefore, from our perspective it would be preferable to speak of *image identification strings* instead of (semantic) media descriptions or features.

The following characteristics of content-based visual retrieval may be one major reason why SVM classification does not require content-based features. Human similarity judgement is goal-centred, influenced by many complex factors and therefore almost unpredictable. Hence, it is a very difficult task to represent it adequately by content-based features. For general application, it may be a more reasonable approach to abandon extracting content-based features in favour of maximising the discrimination ability of features. Eventually, content-based visual retrieval shifts from signal processing more and more towards information retrieval. The classifier used becomes the core component of the user-centred retrieval process. Currently, SVMs are the tool of choice for the imitation of human visual similarity perception.

4.2.3 Application of Context-Free Features

Without doubt, these findings make the application of sophisticated signal processing and semantic enrichment in user-centred retrieval partially questionable. However, the positive consequences of querying by classification of image ID strings are extraordinary. Image ID strings can be computed (assigned) quickly and do not have to be stored in a repository. Retrieval becomes a light-weight process. Systems can be developed and debugged significantly easier, since an entire level of sophistication falls away. Eventually, the results for the best context-free features are (despite higher redundancy and other data anomalies) even better than the results for content-based features. Context-free image ID strings seem to provide more space for discrimination (the desired type of semantics). This is a serious advantage in a research domain where it is silently accepted that the human influence (visual similarity judgement) may appear in an almost arbitrary fashion.

Furthermore, image ID strings allow for real-time feature extraction. Actually, it is sufficient to assign media objects with image ID strings from a pre-defined catalogue. Even this association has to be performed only once on the first retrieval iteration. The resulting retrieval process is light-weighted and can be performed ad hoc. All sophistication is encapsulated in the classifier. Hence, a retrieval situation is fully described by the Lagrange coefficients of the trained SVM. For frequently appearing queries it could make sense to store these coefficients as a *semantic feature* (the additional knowledge is provided by the user's relevance judgement). Moreover, features that can be constructed in real-time allow for the implementation of flexible multimedia database management technologies such as the mediator paradigm [20]. If content similarity is sufficiently described by human feedback encapsulated in straightforward classifiers, it can be handled human-like flexibly.

However, some open problems remain. Content-based visual retrieval is often implemented as a two-step process. In the first step, distance measures and content-based features are used to retrieve a first result set. This set is iteratively refined by relevance feedback in the second step. Obviously, image ID strings cannot be employed for distance-based querying in metric spaces (e.g. by following the vector space model). Therefore, we suggest to replace the first step by a randomly chosen selection of media objects. If available, domain knowledge can be used as well.

4.2.4 Summary

SVM-based classification of users' relevance judgement is – especially on smaller collections – a hard to beat retrieval method. In fact, SVMs perform so well, it hardly matters *what* is classified. From the experiments in this case study we can see that simple context-free features perform as well (or even better) as sophisticated signal processing operations, for example, the content-based visual MPEG-7 features. SVMs require features only as an adequate *ground for discrimination*. Hence in this context, we consider it more precise to speak of image ID strings instead of features. The best image ID string identified in the experiments is random-based. However, from the insights of the data analysis section we understand that a uniform random function would not provide enough space for discrimination. Therefore, the random function is augmented by a simple step function. This construction outperforms the MPEG-7 features. These surprising findings show the *crucial power of the retrieval step*. Well-discriminating features can be computed in real-time during the retrieval process. Image retrieval turns into a more flexible process that is exclusively based on human visual similarity perception.

5 Conclusions

This contribution intends to:

- make the meaning and structure of content-based features transparent and explain what semantic enrichment means for the structure and content of visual features.
- explain in simple words the frequently used methods for feature extraction, semantic enrichment and cyclic retrieval.

- make the reader familiar with techniques for statistical analysis of feature data and explain what can be expected from content-based features.
- uncover the partial contradiction in the requirements of content-based features that should, at the same time, summarise the media content adequately and provide an efficient ground for discrimination.

One major conclusion of this contribution is that modern classification methods are such powerful retrieval methods; they can classify basically *any* kind of data efficiently. We believe that – compared to the users' relevance input (on media objects) – the shapes of the data vectors employed for media description (randomly chosen, semantically enriched, etc.) and their data quality are only of minor importance for the quality of retrieval results.

References

1. Breiteneder C, Eidenberger H (1999) Content-based image retrieval of coats of arms. Proc. of IEEE Multimedia Signal Processing Workshop, pp 91-96
2. Del Bimbo A (1999) Visual information retrieval. Morgan Kaufmann
3. Chang SF, Sikora T, Puri A (2001) Overview of the MPEG-7 standard. IEEE Transactions on Circuits and Systems for Video Technology 11/6: 688-695
4. Eidenberger H, Breiteneder C (2002) Macro-Level Similarity Measurement in VizIR. Proc. of IEEE ICME, Lausanne, Switzerland, pp 721-724
5. Eidenberger H (2004) Statistical analysis of the MPEG-7 image descriptors. ACM Springer Multimedia Systems Journal 10/2: 84-97
6. Fuhr N (2001) Information Retrieval Methods for Multimedia Objects. In: Veltkamp RC, Burkhardt H, Kriegel HP (eds) State-of-the-Art in Content-Based Image and Video Retrieval. Kluwer, Boston, pp 191-212
7. Hough PVC (1962) A Method and Means for Recognizing Complex Patterns. US Patent 3,069,654
8. International Standards Organization (2002) MPEG-7 Information Technology – Multimedia Content Description Interface – Part 3: Visual. ISO/IEC 15938-3:2002(E)
9. Jain AK, Murty MN, Flynn PJ (1999) Data Clustering: A Review. ACM Computing Surveys 31/3: 264-323
10. Joachims T (last visited 2007-08-01) SVM light. svmlight.joachims.org
11. Kohonen T (1990) The self-organizing map. IEEE Proc 78/9: 1464-1480
12. Lew MS (2001) Principles of visual information retrieval. Springer, Berlin
13. Loehlin JC (1998) Latent variable models: An introduction to factor, path, and structural analysis. Lawrence Erlbaum Assoc, Mahwah, NJ
14. Mallat SG (1989) A Theory of Multi-Resolution Signal Decomposition: The Wavelet Representation. IEEEPAMI 11: 674-693
15. Manjunath BS, Ohm JR, Vasudevan VV, Yamada A (2001) Color and texture descriptors. IEEE CSVT 11/6: 703-715

16. Manjunath BS, Salembier P, Sikora T (2002) Introduction to MPEG-7. Wiley
17. Marques O, Furht B (2002) CB image and video retrieval. Kluwer, Boston
18. Müller KR, Mika S, Rätsch G, Tsuda K, Schölkopf B (2001) An Introduction to Kernel-based Learning Algorithms. IEEE TANN 12/2: 181-202
19. Rao KR, Yip P (1990) Discrete Cosine Transform: Algorithms, Advantages, Applications. Academic Press, Boston
20. Santini S, Gupta A (2004) Mediating imaging data in a distributed system. Proc. of SPIE Storage and Retrieval Methods and Applications for Multimedia Conference, San Jose, CA, pp 365-376
21. Sanz JLC, Hinkle EB, Jain AK (1998) Radon and Projection Transform-Based Computer Vision. Springer, Berlin
22. Schäfer G, Stich M (2004) UCID - An uncompressed colour image database. Proc. of SPIE Storage and Retrieval Methods and Applications for Multimedia, Conference San Jose, CA, pp 472-480
23. TU Munich (last visited 2007-08-01) MPEG-7 experimentation model. www.lis.e-technik.tu-muenchen.de/research/bv/topics/mmdb/e_mpeg7 html
24. University of California Berkeley (last visited 2007-08-01) Corel dataset website. elib.cs.berkeley.edu/photos/corel/
25. Vapnik VN (1995) The nature of statistical learning theory. Springer, Berlin

User-Centered Multimedia Retrieval Evaluation based on Empirical Research

Mathias Lux[1], Gisela Granitzer[2], and Günter Beham[2]

[1] Department for Information Technology (ITEC), Klagenfurt University,
Universittsstrasse 65-67, 9020 Klagenfurt, Austria `mlux@itec.uni-klu.ac.at`
[2] Know-Center, Inffeldgasse 21a, 8010 Graz, Austria
`{ggrani,gbeham}@know-center.at`

The evaluation of retrieval mechanisms for inter-method comparison is necessary in academic as well as in applied research. Major issue in evaluations is in which way and to what extent the actual perception of users from the target user group is integrated. Within multimedia retrieval systems impressions and perceptions of users vary much more than in text retrieval. Empirical studies are a common tool in social science and offer a way to research the correlation between the user perception and the computed similarity between pairs of multimedia documents or a query and the set of results. This approach can be used to complement and extend current evaluation approaches. Within this contribution we revisit and summarize general methods from social science and psychology for the interested reader in the area of computer science with some knowledge about statistics. Furthermore we give two examples of undertaken empirical experiments and their outcomes. Within the first one the perception of users is investigated and compared to factors like educational context and gender, while in the second study metrics are tested upon their ability to reflect the notion of similarity of users. Both experiments aim to give examples and insight on how empirical studies can be conducted in multimedia research in general and multimedia retrieval evaluation in special.

1 Introduction

In computer science, which derives a lot of research methods and traditions from mathematics, it is generally assumed that computers operate on logical levels, where true and false can be distinguished clearly. Research areas and trends like fuzzy computing or genetic programming already drift away conceptually from this assumption. In typical information systems users usually interact with the system by for instance searching or creating content. As soon as user interaction, becomes part of a process, true and false are concepts, which might not be appropriate any more for the situation. An even

M. Lux et al.: *User-Centered Multimedia Retrieval Evaluation based on Empirical Research*, Studies in Computational Intelligence (SCI) **101**, 175–194 (2008)
www.springerlink.com

more complicated issue is to deal with information that algorithms and programs cannot understand and interpret (yet), like video streams or digital photos.

In multimedia retrieval many different retrieval methods have been developed over the last years. The importance of multimedia retrieval has been identified for instance in [9]. In this publication the authors also outline the problem that multimedia retrieval evaluation is a far more challenging task than the evaluation of text retrieval mechanisms. One major problem is the availability of generally agreed and comprehensive standardized test data sets like in text retrieval evaluation.

Evaluation methods from the area of text retrieval were adopted for multimedia retrieval by several retrieval evaluation initiatives (see section 2 for an overview). Unfortunately they focus on very specific topics and the test sets cannot be applied in arbitrary scenarios. Furthermore the significance of information retrieval evaluation is under discussion in the text retrieval community: In [23] for instance the notion of significance of information retrieval evaluation results is discussed and error rates for the TREC test collection are presented.

In many cases no existing test set satisfies the requirements for the specific use cases in the respective application domain. If none of the already existing test sets and methods are applicable, two different options can be considered: (i) A test set including topics and results can be created. This is a rather time consuming task as the data set has to be huge and relevance ranking for the topics has to be done by multiple experts. (ii) Alternatively one can use another evaluation method. A promising candidate for the evaluation of multimedia retrieval mechanisms is the method of the *empirical study* . In contrast to classical retrieval evaluation measures, which try to eliminate user dependencies to achieve an arbitrarily usable test sets, actual users are integrated in empirical studies. Empirical studies cannot yield fully true (or false) conclusions. But with the tool of empirical studies conclusions at certain probability levels can be drawn. Within this chapter we revisit selected common methodologies for empirical research from social science. We put the methods into context of multimedia research and present example studies.

The contribution is structured as follows: The introduction is followed by a brief section introducing related aspects and work. After the introductory words a literature survey on empirical studies is given. Following this theoretical part two different example studies are presented. The first one aims at the evaluation of metrics for the retrieval of digital photos and the second aims at discovering insights on the perception of different (user) groups. Finally, this contribution is concluded and summarized.

2 Related Work

Evaluation of multimedia retrieval algorithms generally relies, like in text retrieval, on huge data sets containing documents, sample queries (or topics) and relevance judgements. Based on the test data set the effectiveness of a system is assessed by measuring the set and ranks of relevant documents a system is able to find. Measures like *relevance* and *recall* (see e.g. [1], [27]) allow then a comparison of retrieval methods. Note that in this approach the actual perception and subjective interpretation of target users is not part of the evaluation methodology. It is solely reflected by the relevance judgements, which were provide by experts and are part of the test collection.

Different evaluation forums and initiatives offer platforms and test collections focusing on different retrieval tasks: In the video track of the Text Retrieval Conference[3] (TRECVID), a forum for the evaluation of video retrieval algorithms, focus is put on shot detection, high level feature extraction and search based on topics. The topic based approach as well as the high level feature extraction are also used by the Cross Language Evaluation Forum (CLEF) in the imageCLEF[4] track. In imageCLEF the collection consists of – as the name indicates – images instead of videos and emphasis is put on cross language retrieval of text associated with the images. The Initiative for the Evaluation of XML Retrieval (INEX) multimedia track[5] also focuses on images. The collection is based XML descriptions of images and images referenced in XML descriptions.

In human computer interaction (HCI) research statistical methods are a common tool. Many research groups use statistics for the evaluation of user perception as well as the accuracy of algorithms compared to the view of the actual user. For instance Rodden et al. in [22] assess the subjective practicability of content based image organization. Tilinger and Sik-Lanyi [26] investigate the difference between left handed and right handed users in navigation in 3D environments. Unfortunately in many published research results questionnaires and captured samples are interpreted subjectively. Furthermore in many cases hypotheses and research questions are also not formulated systematically. Although matching the requirements of common HCI evaluations employed methods can be further enhanced to match the standards of empirical studies.

In the ITU recommendation ITU-R BT.500-11 "Methodology for the subjective assessment of the quality of television pictures" (see [13]) methods for the evaluation of subjective quality of television pictures are standardized. The recommendation includes instructions for testing and analysis. Furthermore it describes methods for calculation of significance of found medians and how studies are planned and undertaken. Therefore it provides the reliability

---

[3] URI: http://www-nlpir.nist.gov/projects/trecvid/
[4] URI: http://ir.shef.ac.uk/imageclef/2006/
[5] URI: http://inex.is.informatik.uni-duisburg.de/2006/mmtrack.html

of a well planned empirical study in the domain of television image quality assessment. Although the presented approach is very well described the provided instructions in the recommendation are not generic enough to adopt them to other fields.

3 Methodological Approach

As the area of statistics and empirical studies is very broad, we can only highlight most important terms and steps to be considered when planning and conducting empirical studies and significance tests. The introduction is mainly based on work and findings presented in [4], [3], [7], [11], [20] and [28], where the interested reader can find additional and detailed information. We consider an introduction to empirical research in connection to computer sciences important, since it is an often overlooked but very helpful tool for supporting ones findings statistically.

3.1 Foundations of Empirical Studies

The actual core of any empirical study are *research questions*, which define what the study actually investigates. From the research questions so called *empirical hypotheses* are derived, which are preliminary answers to the research questions. They specify expectations concerning certain facts.

Empirical hypotheses are translated into *statistical hypotheses*, which represent them in the form of statistical units and their value. Hypotheses can be differentiated into null hypotheses H_0 and alternative hypotheses H_A where null hypotheses usually are the ones intended to be rejected in favour of the alternative hypotheses. The aim is to statistically reject the null hypothesis and therefore support the empirical hypothesis. Note that the empirical hypotheses cannot be proven but only retained with statistical means. To give an example, an empirical hypothesis H_0 could say that men and women differ in average reaction time for deciding about picture similarity. The statistical hypothesis H_0 would be $\mu_1 - \mu_2 = 0$ (μ_1 and μ_2 denote the avergae reaction time of men and women), stating that the samples are drawn from populations whose parameters μ_1 and μ_2 are identical.

Concerning the range of hypotheses we distinguish *singular, existential, and universal hypotheses* :

In general quasi universal hypotheses are used, which refer to a restricted Population. Depending on the kind of expectations expressed by the hypothesis it is differentiated between *correlation, differential, and change hypotheses*, which might be *directional* or *non-directional*. Correlation hypotheses state a covariation between variables, differential hypotheses state that groups of subjects or objects differ regarding a certain variable, and change hypotheses state the changing of a variable over time. Directional hypotheses state that for example reaction time is shorter for men than for women, whereas a

Table 1. Possible range of hypothesis

Hypothesis range	Hypothesis holds for ...
singular	one single subject of the population
existential	at least one subject of the population
universal	every single subject of the population

non-directional hypothesis states that there is a difference, not specifying the direction.

A hypothesis qualified for research has to fulfil certain *criteria*. It has to be:

- consistent – hypotheses must not contain contradictions in their logical assembly
- criticisable – it has to allow for observations, which might falsify the hypothesis, and
- operationalisable – it has to be assured that the terms used in the hypothesis can be assigned observable phenomena, namely empirically observable indicators.

There are *different kinds of empirical studies*: it is differentiated between laboratory and field investigations, and experiments and quasi experiments. In a laboratory investigation confounds, which are interfering variables, related to the investigation itself such as noisiness are controlled, which is not the case in a field investigation. In an experiment confounds related to the subjects, such as motivation, are controlled. This is not the case in a quasi experiment. In any kind of investigation at least one variable, called *independent variable* , is systematically varied, and the effect of variation is measured by observing the changes in a dependent variable. The concept of confounds, independent and dependent variables will be further deteiled in the subchapter following.

Following illustrates the concepts discussed so far: Findings in multimedia retrieval might suggest that the pace, at which humans judge visual similarities, depends on the interest of a person in a topic. One possible research question is: Does the interest in a topic determine the pace at which visual similarity judgments are executed? For a corresponding empirical hypothesis, the directional alternative hypothesis H_A, would say: There is a positive correlation between interest in a topic and the pace of judging the visual similarity of topic specific pictures. The respective null hypothesis H_0, which we aim to reject in order to support H_A, is: There is no positive correlation or no correlation at all between interest in a topic and the pace of judging the visual similarity of topic specific pictures. Since we are working with correlations the statistical hypothesis would be $\rho > 0$. Here we have universal correlation hypotheses, since a covariation is postulated for all cases and no restriction is made concerning the population. These hypotheses also fulfil the criteria of being consistent, criticisable, and operationalisable, since there are ways of determining or assessing interest, judgement pace, and subjective visual similarity.

3.2 Experimental Design & Procedure

The *experimental design* refers to the logical set up of the study, which allows testing of the hypotheses in order to reject or support them.

The experimental design is essentially determined by the *independent* and *dependent variables*. Independent variables are varied corresponding to the specifications of the variable. The dependent variables are observed whether the variation has an effect. To give an example, the independent variable gender would have the specification man and woman. One might expect the indeopendent variable "gender" has an effect on the dependent variable "reaction time in judging picture similarity". Depending on number of independent and dependent variables we refer to the *experimental design* as *one or more factorial* and *one or more variate* experimental design. The specifications of the independent variables and their combinations respectively determine the *experimental conditions*. Considering two independent variables like "gender" (with specification "man" or "woman") and "educational background" (with specification "technical education" and "non-technical education") an example experimental condition is "woman with technical education".

Variables, which systematically distort the relationship between independent and dependent variables, are called *confounds*. These have to be controlled, eliminated or held constant. Subjects or objects, in which the dependent variable is observed, are assigned to the experimental conditions. Usually one can distinguish between *experimental* and *control groups*, where the latter serves as a reference: The experimental group undergoes a certain treatment that is experimental condition, while the control group does not. Depending on kind and complexity of the test arrangement – pre-tests can be performed, one or more groups can be studies, repeating measurements can be done, etc. – different experimental designs can be distinguished. Due to space retrictions these are not discussed here. Refer for instance to [7] or [4] for more information.

When conducting an empirical study a *sample* has to be drawn. The sample may consist of human subjects or objects obeying the defined *inclusion* or *exclusion* criteria. So, subjects might be required to have or lack certain characteristics such as a certain type of education. One can distinguish between different kinds of samples: A *random sample*– each unit of the population has the same chance of being selected – allows for a good generalisation, but especially true random samples are very difficult to obtain. *Convenience samples* are more usual, they are drawn at the convenience of the researcher and the availability of subjects or objects. Of course, in this case generalisation of the findings of the study is restricted (see e.g. [28]). Next to the purpose of the study and the population, the *sample size* is determined by meeting constraints based on the following factors:

- The *level of precision*, which is the range, within we expect the true value of the population to be located. The sample has to be big enough to guarantee a certain precision.
- The *confidence level*, which indicates the certainty, the observed value will lie within the range of precision. The sample size has to be large enough to guarantee a certain confidence level.
- The *degree of variability*, which concerns the distribution of an attribute in the population. The larger the variability, the larger the sample size is required for reaching a given level of precision.

3.3 Data Assessment Methods

Several methods and techniques for data assessment are described in literature. Common examples are *questionnaires, study of behaviour*, or *controlled presentation of stimuli*. Observations of any kind are translated into *data*, which might adopt different *scale levels*. One can differentiate between nominal, ordinal, interval and ratio scales for measurement. A nominal scale allows assertions concerning equality and inequality, an ordinal scale about larger/smaller relations, an interval scale about the equality of differences, and a ratio scale allows assertions about the equality of ratios:

- *nominal*: classification using symbols, e.g. male & female
- *ordinal*: symbols with a pairwise relation, e.g. ordering, for instance high, medium and low confidence concerning judgements
- *interval*: accurate distance between values can be calculated, e.g. numerical values between 1 strongly agree and 10 strongly disagree
- *ratio*: a meaningful zero point is available, e.g. time scale or weight scale

Note that various statistical tests require certain scales and data properties such as for instance a normal distribution of assessed numerical values. In that sense the *kind of data* and the planned *statistical tests* have to be determined a priori.

When the investigation is carried out, it is important to care for standardisation of the applied methods to allow comparability and the communication of conditions to the readers to allow reproduction of the study results. Furthermore each subject or object of investigation must find oneself within the experiment in the same situation only differing by experimental condition. Depending on the matter of investigation single or group testing is possible. Experimenter biases also have to be eliminated or controlled: The instruction part for instance, where the subjects of investigation are briefed in advance of the tests, is very important: Subjects must fully understand what is expected from them and they must be introduced to the experiment always in the very same way.

3.4 Evaluating the Data

Although lots of different methods for data analysis exist and certainly a lot can be said about statistical analysis and statistical test theory, but in this section only basic statistical procedures also used in the example studies in section 4 and section 5 are summarized and presented. This of course should not belie the necessity of carefully studying the literature on statistical methods and selecting those which are most appropriate for the given situation. So, parametric tests for example (*what is a parametric test*) require data on interval scale, normally distributed data and homogeneous variances. If these conditions are not fulfilled a non parametric test would be more appropriate.

The most common and widely used technique is the *t-test*, which is also used in our second example study (see section 5) to find a significant difference between the mean of two variables. The t-test is used to determine whether a differential null hypothesis is rejected or retained. It is for instance applied to test expected values based on a population following a normal distribution, with the same variance. The expected values, for instance the mean, are transformed into a test value t, which follows a student-t distribution for a valid hypothesis H_0. Using the student-t distribution the probability of validity of H_0 based on t can be calculated, which gives the significance level (see e.g. [11] and [4]).

Principal component analysis(PCA) is a numerical method for reducing the number of dimensions of a data set. Used in statistics it is applied to a combination of group related variables to form more general variables (factors) describing a previously assessed dataset. More detailed information about PCA can be found in [15]. PCA is used in the second example to test whether factors of the conducted data correspond to the variables defined in the hypothesis.

The contiguity between two variables is statistically represented by the *correlation coefficient* ρ. The correlation coefficient results from standardising the covariance, which reflects the extent of the linear association between two variables. It can adopt values from $[-1, 1]$: -1 indicates a perfect negative correlation, 1 indicates a perfect positive correlation, and 0 indicates no correlation (see for instance [11] or [5]). To give an example, a negative correlation between interest in a topic and the pace of judging the visual similarity of topic specific pictures means that a high interest in a topic goes along with a slow reaction time. Depending on the scale levels of the variables different correlation coefficients can be calculated. So for example the point biserial correlation coefficient is based on dichotomous and interval scaled variables, a product moment correlation coefficient is based on two interval scaled variables and a rank correlation, like Spearmans ρ, is used for instance for two ordinal scaled variables. For a statistical validation a bivariate normal distribution is required. In this case ρ is a good estimator of the contiguity. If this condition is not fulfilled the estimate is out by $1/n$ but with an increasing n the inaccuracy is negligible. If an empirical correlation proves the null hypothesis, namely whether $\rho = 0$, can be decided by a t-test. If the calculated t is larger than the critical t, which can be read from corresponding tables – its size depends on the determined significance level and whether the test is one or two sided – the correlation is significant. To investigate whether two correlation coefficients differ significantly the correlation coefficients are transformed into Fischer z values and z value difference is calculated. The probability corresponding to z can be read from given tables in the literature or can be computed with common statistical software. If it is equal to or smaller than 0.05, representing a significance level of 5%, which is an overall agreed border for significant results, the coefficients differ significantly from each other. If it is equal to or smaller than 0.01 the difference between the coefficients is strongly significant, representing a significance level of 1%, which is an generally agreed border for strongly significant results.

3.5 Interpretation

The interpretation of results completes an empirical study. The interpretation of the data has to be done against the background of the sample, the experimental design, sample size, and used statistical tests. Interpretation also includes pointing at constraints and suggesting for improvements (see also [28] for common flaws and further guidelines).

4 Example A: Most Semantic Metrics for Retrieval

The first example gives a classical approach of user centered evaluation. Several metrics and parameter sets for metrics for multimedia retrieval are given. The task is to select a metric, which manages to reflect the notion of similarity in this domain.

4.1 Research Question

In general a combination of different low level and high level features is used within an image retrieval system to find images matching user needs. Low level features can be extracted from multimedia documents automatially. They are represented mostly by numeric values and vectors describing characteristics of an image in a way that is (i) useful and (ii) efficient enough for retrieval. Examples for those low level features are colour histograms, dominant colours of images and texture characteristics like regularity or coarseness. High level features on the other hand cannot be extracted without additional manual input (see [24], [8]).

Prominent problems in multimedia retrieval are (i) the *selection of appropriate features* (see e.g. [19]) and (ii) the *selection of appropriate metrics for the features* (see [25]). Imagine for instance a company, where web sites and advertisements are designed. Employees search for royalty free images, which contain a certain amount of a colour or have a certain texture. For this use case colour and texture based low level features fit perfectly. A medical research team on the other hand might search for X-Rays containing broken ribs. In this case meaningful search results based on colour and texture based low level descriptors are unlikely. For high level metadata also multiple methods for retrieval are available. One can for instance select from various weighting schemes like TF*IDF and BM-25 (see [1], [21]) and from different methods for word stemming and disambiguation. Through combination and parameterization of different methods various metrics for high level metadata can be defined. Therefore the research question arises: Which metric fits best?

Based on the aim to find the best fitting metric out of the set of metrics M following hypothesis H_A was formulated: There is a strong correlation between the test metric $m_i \in M$ and the perception of the user. The respective null hypothesis H_0 is: There is no correlation between the metric $m_i \in M$ and the perception of the user.

4.2 Materials

In our use case the test data set contained 96 photos and associated metadata. The metadata is provided in the standardized MPEG-7 multimedia description format. In our study we focused on metadata encoded in the *Semantic Description Scheme* , which allows a graph based description similar to RDF, was used for comparison of the images (see [14] for MPEG-7 and the Semantic Description Scheme). 5 different types of metrics were used to compare the description graphs pair wise:

- The *Maximum Common Subgraph* metric from [6], where the Jaccard coefficient of the node sets is used as similarity measure
- The *Error Correcting Subgraph Isomorphism* metric as described in [2], where optimal approximate mapping between two graphs is the foudation of a distance measure
- A text based metric using the textual descriptions in description graphs omitting the structural information (cosine coefficient on term vectors using TF*IDF and BM-25 weighting)
- A generalized path index metric based on the vector space retrieval model using the cosine coefficient and TF*IDF and BM-25 weighting as described in [16]
- A suffix tree and path based metric for comparing labelled graphs as introduced in [16]

Including different tested approaches and parameter settings 122 different variants of metrics were investigated, whereas many more other variants have been tested but not included in the documentation. As the study only focuses on the metrics results details on the actual implementation of the metrics are unneccasary for the description of the evaluation method. However interested readers will find additional information in [17], [18] and [16]. These references describe in detail metrics used in this evaluation as well as the test data set.

The number of samples used for the study depends on the needed significance and the intensity of correlation addressed in H_A. In general a strong correlation is defined by a correlation coefficient of $\rho >= 0.5$. Default constraints for a significance test in research are a significance level of 5% and a statistical power of 80% ($1 - \beta = 0.8$). Based on these numbers 22 samples are needed for the evaluation to yield significant results according to [4]. The number of samples can be further reduced if the effect size is increased. In our experiment we chose a minimum effect size of $\rho = 0.6$ and a used 20 samples. In our case a sample is the similarity values between two images. 20 pairs of images out of 9216 were chosen randomly.

To create an optimal questionaire for the survey a pre test form containing 14 questions was created. Furthermore for each question a slide was created. Each question allowed 5 different answers ranging from "images are very similar" to "images are not similar at all". Summaries of the semantic descriptions were shown on the slides as keywords beneath the pictures. After a pre-test with 14 participants following conclusions for the questionnaire and the slide show were drawn: (i) A presentation time of 20 seconds per slide is appropriate for all users to rate the similarity of the images. And (ii) according to participants the set of keywords is not appropriate to reflect meaning and content of the semantic descriptions, which were the actual input for the similarity metrics.

Based on these findings the slideshow was adapted as shown in figure 1: Full sentences instead of keywords are used to describe the semantics of each image. The questionnaire as well as the slideshow were extended to the number of 20 samples and a second pre-test with 13 participants (others than the ones from the first test) has been undertaken. Within this test a sample was identified, where names had to be disambiguated. As there were two people with the same name (only surnames were used to ensure anonymity of the people shown on the images) within the test set, many users were uncertain if two different people or the same person was described.

4.3 Subjects

To model the perception of the user a group of potential users had to be surveyed. As already mentioned the optimal approach is to select the participants of the survey randomly from the target users. In our case we used a convenience sample: The students of a computer science lecture at Graz University of Technology. The actual survey was undertaken with 112 participants (first time participants, who did not participate within pre-tests).

Fig. 1. Sample slide of the evaluation slide show

4.4 Procedure

Following the two pre-tests described in section 4.2 the actual survey was done: Each participant was given a questionnaire and the prepared slide show was presented. As the survey took place for all participants simultaneously the introduction to the task was the same for each participant.

From the survey results the median values were taken as reference similarity values reflecting the user perception. To ensure the statistical correctness of this approach we ensured with a χ^2-test that the surveyed data follows a normal distribution with a probability $> 99.9\%$.

4.5 Results

A first step in the data analysis was the *visual analysis* of the collected samples. In our scenario we want to investigate a possible correlation between user perception and different metrics. For this task a scatter plot, as shown in figure 2, provides in many cases a deeper insight.

Based on figure 2 the we can draw conclusions for further analysis. The data points in the plot are roughly distributed along a straight line for metric 1, which indicates covariance. Using the correlation coefficient to investigate a contiguity of two variables also requires the variables to be related linearly. Unfortunately plotted data points for metric 2 suggest that the metric might follow steps (several even lines along the x-axis) instead of a straight line. Further analysis of the samples shows that the values of the metrics 1, 5, 6, 7 and 8 of table 2 follow the same distribution as the reference values (verified with a χ^2-test with a probability $> 90\%$). This cannot be assumed for the metrics 2, 3 and 4. In this case a rank correlation coefficient like *Spearmans rank correlation* is a better tool to investigate contiguity between the user perception and the metrics.

Spearmans rank correlation is calculated by converting the actual data into the rank within the sample. Tie ranks are averaged: If for instance 4

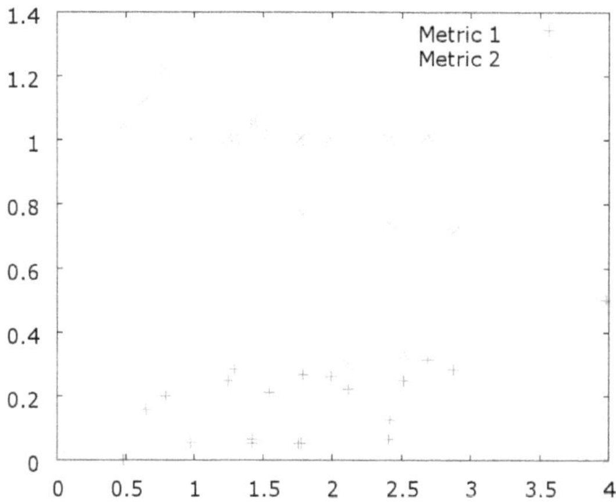

Fig. 2. Scatter plot of 2 different metrics. Reference values are on the x-axis

samples have the same rank (e.g. 10) they are assigned the median of the respective ranks (e.g. $10 + 11 + 12 + 13 = 11.5$). Spearmans rank correlation coefficient ρ is then

$$\rho = 1 - \frac{6 \cdot d_i^2}{n \cdot (n^2 - 1)}$$

whereas n denotes the number of samples and d_i denotes the rank difference of the corresponding values of the two investigated variables. The rank correlation analysis has been applied to the same metrics and the same sample – with the intermediate step of ranking the values – as described in section 4.5.

Table 2. Spearmans ρ for the tested metrics and the reference sample reflecting the user perception

Metric	ρ
1. MCS	0.536
2. ECSI VS	−0.721
3. ECSI B	−0.299
4. VS Text	0.415
5. VS BM-25 Triple	0.684
6. VS BM-25	0.664
7. ST IDF Triple	0.658
8. ST IDF	0.720

Based on the table for significance of Spearmans rank correlation coefficient in [10] the H_0 hypothesis can be rejected for the metrics 1, 2 and 5-8 at a significance level of 0.05. For the metrics 2 and 5-8 this can even be done for a significance level of 0.01.

4.6 Interpretation

Based on scatter plot and distribution analysis of the samples we decided to use a rank correlation coefficient for analysis. The results are shown in table 2. We can see that out of the 8 metrics for two (metric 3 and 4) the H_0 hypothesis could not be rejected. For all other metrics H_0 could be rejected and H_A could be retained. In other words these metrics turned out to be a good candidates for multimedia retrieval based on the MPEG-7 Semantic Description Scheme.

Therefore we conclude that the metrics 1, 2 and 5-8 are promising candidates for an image retrieval system. This conclusion is done under the assumption that the survey participants reflect the target user group and the images reflect the data set of this use case (see sections 4.2 and 4.4). However a best candidate cannot be assumed. Although the values of the computed correlation coefficients intuitively give a relative ranking of metrics, no significant difference could be supported.

The method applied in the experiment is – although quite laborious and time consuming in the survey & data gathering step – a reasonable alternative to evaluation techniques using gold standards and evaluation measures like relevance and recall (see e.g. [1] or [27]) as the number of test documents can be reduced. The novelty in this multimedia retrieval evaluation experiment is the usage of standards in empirical research (see [28]), which ensure the possibility of comparison between projects and research groups, as well as the integration of the target user group in the evaluation.

5 Example B: Perception & Understanding of Groups

Example B is based on the previous example in the sense that its background is also multimedia retrieval and results of conducted pre-tests were used as a basis for preparing materials. The main difference, however, is that Example B focuses more on differences between groups of users and how they perceive similarity of photos.

5.1 Research Question

Past research showed that, on average, differences of cognitive abilities between males and females exist (see [12]). Males benefit from faster manipulation of visual information in the working memory. Females on the other hand score higher when accessing information in the long term memory and achieve a higher perceptual speed. For designing multimedia retrieval it is among other issues very important to know which kinds of influences may have an impact on the perception of image similarity. Questions like the following ones arise in this context: Do gender differences as mentioned before influence the perception of image similarity? Does additional textual information commenting images change the judgement of similarity? Could different backgrounds of users lead to different results?

For this example we defined background as the affiliation to a technical or non-technical university. The rationale behind the distinction of the university background is that people with a more technical background may tend to approach and solve problems differently than those with a non-technical background. Surveying these information could also lead to insights of possible perceptional interferences. Summarising, this example study attempts to find answers whether gender, additional information commenting photos and the university background have influences on the perception of similarity.

After stating the background and the purpose of the study, a hypothesis H_A was constructed out of the stated questions and reads as follows: Users perceive the similarity of digital photos without additional information, digital photos with some keywords, and digital photos with textual description, based on gender and association with a technical or non-technical university, differently.

Taking this formulated hypothesis two different groups of variables can be extracted as already stated in section 3.1. The first group consists of three independent nominal variables with two and three levels, respectively:

- Gender, two levels: (1) male, (2) female
- Association with an university, two levels: (1) University of Graz, (2) Graz University of Technology
- Additional information commenting photos, three levels: (1) pairs of digital photos without additional information, (2) pairs of digital photos with 3 to 6 keywords, (3) pairs of digital photos with a textual description

The second group contains one dependent ordinal variable:

- Similarity judgements

After formulating the hypothesis and extracting variables to be surveyed, the next steps are to describe for instance which materials will be used, how subjects are selected and how the procedure looks like to conduct the data.

5.2 Materials

The first step was to select pairs of digital photos showing more or less apparent differences in low and high level features. For this study four different pairs of images conducted from different photo databases were assembled. Based on the independent variable "Additional information to photos" all four pairs of photos appear three times in the questionnaire, each of them attached with different amounts of additional information. Figure 3 shows one pair of digital photos with keywords attached to them. Each of these pairs was printed on a separate sheet of paper and assembled in a folder. The folder was selected to allow subjects fast browsing through these pages in a short period of time. The pages were also sorted in a way that no two identical pairs followed each other.

This folder was then presented to subjects stating the following question to assess the similarity of a pair of photos: "To what extent do you think these two photos are similar?" It was directly followed by a 6-point rating scale ranging from 1 (not similar) to 6 (very similar). This scale forced the subject to decide whether the presented images were similar to some extent or not. Due to the fact that an even number of choices was offered, a neutral answer was not possible. The idea behind the even-numbered choices is to force subjects to explicitly state wether they think that the presented photos are similar to some extend or not.

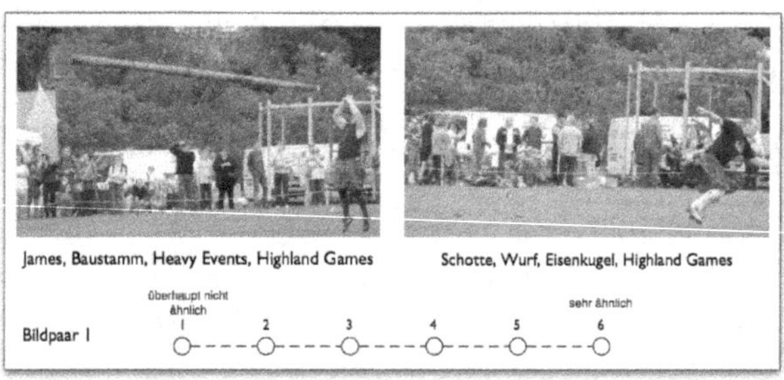

Fig. 3. Sample question of the questionnaire folder

5.3 Subjects

The subjects were 14 female and 14 male students. At the time the study was conducted one half of the subjects were students at the University of Graz and the other half students at Graz University of Technology, respectively. Subjects were only chosen if they could spend at least 10 minutes without getting in a hurry. To achieve some diversity of students data was collected at different places around the university campuses.

5.4 Procedure

The experimenter visited different places at both university campuses and randomly asked students to participate in the study. Each subject was explained that s/he was meant to assess the similarity of pairs of photos. Questionnaire instructions emphasised that each pair of images had to be assessed in at most 20 seconds based on the results gathered by pre test for example A described in Section 4.4. The experimenter took care that pages were turned in time.

5.5 Results

To check the distribution of the assessed data the Kolmogorov-Smirnov test (K-S test) was taken. This test can be used to answer the question whether data is normally distributed or not. Compared to the χ^2-test the KS-test is also suitable for small sets of samples. For all statistical evaluations a significance level of 5% was assumed. Results of the K-S test in table 3 for each pair of digital images (PI) showed that all calculated significance levels are above the critical border of 5% and thus data is normally distributed. Satisfying the precondition of normally distributed data a factor analysis could be calculated.

Table 3. Results of the KS-test

	PI 1	PI 2	PI 3	PI 4
Two-tailed significance levels	0.205	0.156	0.158	0.153

The factor analysis was done as a principal component analysis testing whether the independent variables "additional information commenting photos" extracted from the hypothesis were measuring the defined criteria. Results showed the existence of four different factors and not three as assumed previously. Trying to figure out how to describe what these four factors were measuring, photos belonging to each of the four factor had to be investigated. Investigations uncovered no obvious criteria to describe these factors. The third statistical method was aimed to find significant differences within the assessed data sets. The hypothesis states that there exist differences in the perception of similarity between male and female as well as between different university backgrounds. For this example the significance of mean differences (female and male subjects) were calculated. Results of the t-test showed no significant differences between the data sets of gender and data sets of university background except for one pair of images ($\rho = 0.001$, $M_male = 4.29$, $M_female = 5.50$). Female students assessed this pair of digital images more similar than male students.

5.6 Interpretation

These results lead to the conclusion that the interference of different amounts of information commenting photos with the perception of similarity could not be supported with the formulated hypothesis, the selected pairs of photos and the chosen sample. Results do not allow to state about additional information commenting photos gender or university background and their influences on perception of similarity.

Note that this study has *not* shown that there are no differences in perception. Only H_A has been discarded as H_0 could not rejected. Although the presented test of the user perception and understanding of image similarity in such a limited way (limited in terms of the sample compared to a global population of users) cannot lead to global understanding of user perception and semantics, it can be applied in communities and domains, were similarity assessment multimedia data is needed as for instance in multimedia retrieval. A common example is the retrieval of photos by architectural characteristics. The architects using the system will have their own understanding of similarity of pictures. In a naive approach photos showing similar buildings from similar angles with similar colours are interpreted as similar. In an architectural use case the age of the building, certain details like windows, doors and ornaments or periods, for example Gothic or Romanesque, are more important features for similarity.

6 Conclusions

Empirical research is a powerful tool. Keep in mind that a hypothesis can not be proven, but only supported. Furthermore experiments are labour intensive: In many cases they need pre-experiments and some times do not yield the desired results. Furthermore quantitative studies heavily depend on the questions asked and the way the questions are asked (as already mentioned as experimenter bias in section 3.2). Therefore whole chapters of books describe techniques to set up questionnaires and interview scenarios and testing environments.

In our opinion the method of empirical studies has great value for multimedia retrieval: Instead of evaluating independently from users, the target user group can be integrated. Based on observations and heuristics a hypothesis upon user groups, metrics and parameters can be stated and eventually supported based on the results of a study. However the most important point is that the actual users are an integral part in the evaluation. In domain specific search engines, like retrieval of technical drawings, 3D models or sport scenes, where subjective similarity differs from domain to domain, it is common practice that developers, who are no domain experts, plan, implement and adjust the search engines. With empirical studies the accuracy of such search engines can be evaluated in a way, that integrates the actual users, and the subjective similarity can be "captured".

However empirical studies cannot replace classical evaluation strategies. To allow cross search engine and cross domain evaluation test data sets are a crucial tool.

6.1 Checklist for Empirical Research

Following list summarises the steps that have to be considered when conducting an empirical study. Although this list is neither exhaustive nor complete the most central aspects are listed.

1. Determine the subject matter of your empirical study
   a) Define research question and derive empirical hypotheses
   b) Translate empirical into statistical hypotheses
2. Plan your empirical study
   a) Determine independent variables and their factor levels
   b) Determine dependent variables and determine confounds
   c) Decide about investigation alternatives
   d) Determine the experimental design
   e) Carefully select your sample
   f) Select significance tests for the study in beforehand.
   g) Plan the procedure
   h) Check your experimental design for repeatability
3. Conduct your empirical study
4. Analyse obtained data and perform significance tests
5. Interpret your data

And finally: Document everything carefully and completely so that you and others can repeat the empirical study and reproduce results

References

1. Ricardo A. Baeza-Yates and Berthier Ribeiro-Neto. *Modern Information Retrieval.* Addison-Wesley Longman Publishing Co., Inc., 1999.
2. Stefano Berretti, Alberto del Bimbo, and Enrico Vicario. Efficient matching and indexing of graph models in content-based retrieval. *IEEE Transactions on Pattern Analysis and Machine Intelligence*, 23(10):1089–1105, Oct. 2001.
3. J. Bortz. *Statistik.* Springer, 1999.
4. J. Bortz and N. Doering. *Forschungsmethoden und Evaluation.* Springer, 2002.
5. I.N. Bronshtein, K.A. Semendyayev, G. Musiol, and H. Muehlig. *Handbook of Mathematics.* Springer, 2004.
6. Horst Bunke and Kim Shearer. A graph distance metric based on the maximal common subgraph. *Pattern Recognition Letters*, 19(3-4):255–259, 1998.
7. Bruce A. Chadwick, Howard M. Bahr, and Stan L. Albrecht. *Social Science Research Methods.* Prentice Hall, 1984.
8. Alberto Del Bimbo. *Visual Information Retrieval.* Morgan Kaufmann Publishers, San Francisco, June 1999.
9. James Allan et al. Challenges in information retrieval and language modeling: report of a workshop held at the center for intelligent information retrieval, university of massachusetts amherst, september 2002. *SIGIR Forum*, 37(1):31–47, 2003.
10. Michael W. Eysenck. *Psychology: An International Perspective*, chapter Research Methods: Appendices. Psychology Press, 2004.
11. Chava Frankfort-Nachmias and David Nachmias. *Research Methods in the Social Sciences.* St. Martin's Press, 1992.
12. Diane Halpern and Mary LaMay. The smarter sex: A critical review of sex differences in intelligence. *Educational Psychology Review*, 12(2):229–246, June 2000.

13. International Telecommunication Union. Methodology for the subjective assessment of the quality of television pictures. Recommendation, 2002. Recommendation ITU-R BT.500-11.
14. Harald Kosch. *Distributed Multimedia Database Technologies*. CRC Press, Nov. 2003.
15. David Lay. *Linear Algebra and Its Applications*. Addison-Wesley, New York, 2000.
16. Mathias Lux. *Semantische Metadaten - Ein Modell zwischen Metadaten und Ontolgien*. PhD thesis, Graz University of Technology, 2006.
17. Mathias Lux and Michael Granitzer. A fast and simple path index based retrieval approach for graph based semantic descriptions. In Benno Stein and Sven Meier zu Eien, editors, *Proceedings of the Second International Workshop on Text-Based Information Retrieval*, Fachberichte Informatik, pages 29–44, Koblenz, Germany, July 2005. Universitt Koblenz Landau.
18. Mathias Lux, Sven Meyer zu Eissen, and Michael Granitzer. Graph retrieval with the suffix tree model. In *Proceedings of the Workshop on Text-Based Information Retrieval TIR 06*, Trento, Italy, August 2006.
19. D. Mitrovic, M. Zeppelzauer, and H. Eidenberger. Analysis of the data quality of audio features of environmental sounds. *Journal of Universal Knowledge Management (JUKM)*, 1(1):4–17, 2006.
20. H.T. Reis and C.M. Judd, editors. *Handbook of Research Methods in Social and Personality Psychology*. Cambridge University Press, 2000.
21. Stephen Robertson, Hugo Zaragoza, and Michael Taylor. Simple bm25 extension to multiple weighted fields. In *CIKM '04: Proceedings of the thirteenth ACM international conference on Information and knowledge management*, pages 42–49, New York, NY, USA, 2004. ACM Press.
22. Kerry Rodden, Wojciech Basalaj, David Sinclair, and Kenneth Wood. Does organisation by similarity assist image browsing? In *CHI '01: Proceedings of the SIGCHI conference on Human factors in computing systems*, pages 190–197, New York, NY, USA, 2001. ACM Press.
23. Mark Sanderson and Justin Zobel. Information retrieval system evaluation: effort, sensitivity, and reliability. In *SIGIR '05: Proceedings of the 28th annual international ACM SIGIR conference on Research and development in information retrieval*, pages 162–169, New York, NY, USA, 2005. ACM Press.
24. S. Santini and R. Jain. Beyond query by example. In *Multimedia Signal Processing, 1998 IEEE Second Workshop on*, pages 3–8, Redondo Beach, CA, USA, Dec 1998. IEEE.
25. Simone Santini, Amarnath Gupta, and Ramesh Jain. Emergent semantics through interaction in image databases. *IEEE Transactions on Knowledge and Data Engineering*, 13(3):337–351, 2001.
26. Adam Tilinger and Cecilia Sik-Lanyi. *Digital Multimedia Perception and Design*, chapter Issues of Hand Preferences in Computer Presented Information and Virtual Realities, pages 224–242. Idea Group Publishing, 2006.
27. C.J. van Rijsbergen. *Information Retrieval*. Butterworths, 1979.
28. Leland Wilkinson and Task Force on Statistical Inference APA Board of Scientific Affairs. Statistical methods in psychology journals: Guidelines and explanations. *American Psychologist*, 54(8):594–604, August 1999.

Specification of an MPEG-7 Query Format

Mario Döller

Department of Distributed Information Systems, University of Passau, Innstrasse 43, 94032 Passau, Germany
`Mario.Doeller@uni-passau.de`

The ever growing increase of digital media content by commercial content providers as well as by user driven multimedia content production necessitates intelligent mechanisms for efficient navigation, search and retrieval in large distributed multimedia content repositories. Besides the well known multimedia standards (MPEG-1, MPEG-2 and MPEG-4), the MPEG consortium also established a metadata (data about data) standard (called MPEG-7) for describing multimedia content. Currently, this international standard is probably the richest multimedia metadata set available. A lot of research has been accomplished on building MPEG-7 enabled repositories. However, a common query format that enables the user to query multiple distributed metadata databases does not yet exist. Coping with this situation, the MPEG committee decided to instantiate a call for proposal (N8220) for an MPEG-7 query format (MP7QF) and specified a set of requirements (N8219) for the purpose of defining a standardized way to uniformly query MPEG-7 enabled multimedia databases. In addition, one clearly expressed desire of that call was to propose a possible Web Service binding as part of the contribution. This chapter introduces a MP7QF query language and describes the background and requirements as well as the main architectural concepts and associated MP7QF XML schema types. In addition, an overview of the framework components such as session management, service retrieval and its usability is presented. However, the main focus of this chapter is on definitions and explanations of the input and output query format.

1 Introduction

During the last years, a vast variety of multimedia information has been brought to recipients due to the growing abilities of computer, telecommunications and electronic industry. Furthermore, a high amount of digital video and audio information has become publicly available over the years. The traditional TV broadcast business moves quickly into the digital area. By having

M. Döller: *Specification of an MPEG-7 Query Format,* Studies in Computational Intelligence (SCI) **101**, 195–216 (2008)
www.springerlink.com

more and more content digitally available, the ability increased to deliver rich information to customers. But at the same time, the problems of management, delivery and retrieval increased because of the data size and complexity. Descriptive information about digital media which is delivered together with the actual content represents one way to facilitate this search immensely. The aims of so-called metadata ("data about data") are to e.g. detect the genre of a video, specify photo similarity or perform a segmentation on a song, or simply recognize a song by scanning a database for similar metadata. A standard that has been established to specify metadata on audio-visual data is MPEG-7 [16] and has been developed by MPEG. This organization committee also developed the successful standards known as MPEG-1 (1992), MPEG-2 (1994) and MPEG-4 (version 2 in 1999). MPEG-7 had a broad impact to experts of various domains of multimedia research. In recent years, the work on databases has been started, in order to store MPEG-7 descriptions in multimedia databases (e.g., MPEG-7 MMDB [12], PTDOM [33]). As MPEG-7 relies on XML Schema, two basic approaches can be distinguished: (1) native XML databases (e.g., Tamino [23]) and (2) mapping strategies for (object)-relational databases (e.g., Oracle [19]). There are several multimedia applications supporting and using MPEG-7. To mention only a few: VideoAnn [26], Caliph and Emir [15] etc. Furthermore, many document query languages such as XML-QL 99 [2], XQL [21], SVQL [6], recent W3C XQuery [31], etc., have been proposed for XML or MPEG-7 document retrieval. However, these languages are not adequately able to support MPEG-7 description queries for mainly two reasons: (1) often they do not support query types which are specific for retrieving multimedia content such as query by example, query based on spatial-temporal relationships, etc. (2) No standardized interface is defined and each query language or MPEG-7 database offers its own query interface, which prevents clients experiencing aggregated services from various MPEG-7 databases. Therefore, the MPEG standardization committee decided to start the work on a query format based on the metadata standard MPEG-7 (MP7QF). The objective of this MP7QF framework is to provide a standardized interface to MPEG-7 databases allowing the multimedia retrieval by users and client applications, based on a set of precise input parameters for describing the search criteria and a set of output parameters for describing the result sets.

This book chapter analyzes currently available query languages and approaches for multimedia and XML data and shows the way, how it will be done within MPEG. Besides a summarization of components of the framework and its usability, this book chapter concentrates on definitions and explanations of the input and output query format.

The remainder of this chapter is organized as follows: First, in Section 2 a short introduction to the MPEG-7 standard is provided. Next, Section 3 specifies requirements of an MP7QF framework. Then, Section 4 deals with related work to MPEG-7 query languages and databases. In Section 5, the main parts of the proposed MPEG-7 Query Format are introduced whereas

Subsection 5.1 introduces an architecture and Subsection 5.2 presents the respective query language. The current status of the MP7QF Committee Draft is outlined in Section 6 whereas accepted concepts of our proposal are highlighted. Finally, this paper is summarized in Section 7.

2 MPEG-7

MPEG-7 [16, 17] is an ISO/IEC standard developed by MPEG (Moving Picture Experts Group) and formally known as "Multimedia Content Description Interface". This organization committee also developed the successful standards known as MPEG-1 (1992), MPEG-2 (1994) and MPEG-4 (version 1 in 1998 and version 2 in 1999). But, MPEG-7 is the first standard of ISO/IEC that exclusively concentrates on describing multimedia content in a semantically rich manner. The standard is organized in eleven parts (from system, low- and high-level multimedia description schemes, to reference software and conformance) and provides a rich set of standardized tools for describing multimedia content. It is important to notice that MPEG-7 documents are coded with the help of the XML standard following the rules of a Data Definition Language (DDL). The DDL provides means for defining the structure, content and semantics of XML documents.

Fig. 1. Left side: Soccer Game Image, Right side: the corresponding MPEG-7 document

In order to demonstrate the use of MPEG-7 for describing multimedia content, Figure 1 shows an annotation for a soccer image. The image is decomposed into two sub-images, each specified with a *StillRegion* DS. We specified for each sub-image an own id, e.g., midfieldTactics, a *TextAnnotation* that contains text description of the image, and a *VisualDescription* which figures a feature vector of size 64 representing the color distribution of the subregion. The global information is described with an *Image DS* that contains in our example a *MediaLocator* specifying the location and the name of the image and a *SpatialDecomposition DS* indicating the spatial characteristic of all subregions.

3 Requirements for an MPEG7 Query Format

In general, one can distinguish between two main types of requirements: requirements for the framework (e.g., session concept, connection to multiple MP-7 databases, etc.) and requirements for the query language. In the following, selected requirements are explained in more detail. A complete list is given in MPEGs requirement paper (N8219).

3.1 Requirements for the Framework

1.) *Allowing simultaneous search in multiple databases:* The framework should support the distribution of a single query to multiple search engines.
2.) *Querying database capabilities:* The framework should allow the retrieval of database capabilities such as supported media formats, supported MPEG-7 descriptors/ descriptor schemes, supported search criteria, etc.

3.2 Requirements for the Query Language

1.) *Query Types*
    a.) *Query by description:* Here, the query language should provide means for the retrieval based on textual descriptions (e.g., names) as well as the retrieval based on desired MPEG-7 descriptions and/or description schemes.
    b.) *Query by example:* The query language should support the retrieval based on representative examples of the desired content.
    c.) *Multimodal query:* The query language should provide means for combining retrieval operations based on different media types.
    d.) *Spatial-temporal query:* The query language should support retrieval based on spatial and/or temporal relationships (e.g., search for images where a red Ferrari is in front of a white house.)
2.) *Specifying the result set:* The query language should provide means for specifying the structure as well as the content in the sense of desired data types.
3.) *Querying based on user preferences and usage history.*
4.) *Specifying the amount and representation of the result set:* The query language should provide means for limiting the size as well as supporting paging of the result set.

4 Related Work to MPEG-7 Query Languages and Databases

A good overview of the usability of XML databases for MPEG-7 is provided by Westermann and Klas in [32]. The authors investigated among others the following main criteria: *representation of media descriptions* and *access to media descriptions*. To summarize their findings, neither native XML databases (e.g., Tamino [23], Xindice [24]) nor XML database extensions (e.g., Oracle XML DB [19, 20], Monet XML [22], etc.) provide full support for managing MPEG-7 descriptions in respect to their given requirements. Based on identified limitations (e.g., supported indexing facilities for high-dimensional data), the retrieval capabilities for multimedia data are restrictive.

In the following, we take a closer look at available XML and/or MPEG-7 query languages in various approaches and evaluate them corresponding to the requirements presented in Section 3.

XPath [1] (XML Path Language) is a recommendation of the W3C consortium that enables the access to individual parts of data elements in an XML document. In general, an XPath expression consists of a path (where the main difference to filenames or URIs is that each step selects a set of nodes and not a single node) and a possible condition that restricts the solution set. The main disadvantage of XPath expressions is their limited usability in querying XML documents. For instance, it does not provide means for grouping or joins. In addition, XPath on its own provides no means for querying multimedia data in MPEG-7 descriptions based on the presented criteria.

XQuery [31] is a declarative query language and consists of the following primary areas that find their counterparts in SQL. For instance, the *for/let* clauses represent the SQL SELECT and SET statements and are used for defining variables respectively iterating over a sequence of values. The *where* clause complies to the SQL WHERE statement by filtering the selection of the *for* clause. A main part of XQuery is the integration of XPath 2.0 and their functions and axis model which enables the navigation over XML structures. Additional parts provide the ability to define own functions analogous to SQL stored procedures and the handling of namespaces. With regard to our requirements, XQuery does not provide means for querying multiple databases in one request and does not support multimodal or spatial/temporal queries. Nevertheless, there is ongoing work in this direction. For instance, the authors in [34] describe an XQuery extension for MPEG-7 vector-based feature queries. Furthermore, the authors in [7, 6] adapted XQuery for the retrieval of MPEG-7 descriptions based on semantic views. Its adaptation, called Semantic Views Query Language (SVQL) is specialized for retrieving MPEG-7 descriptions in TV news retrieval applications and is not indented to be a general query language. Another adaption of XQuery is presented in [5].

SQL/XML [4] is an extension of SQL and was developed by an informal group of companies, called SQLX[1], including among others IBM, Oracle, Sybase and Microsoft. A final draft has been standardized as SQL part 14 (SQL/XML) by ANSI/ISO in 2003. Its main functionality is the creation of XML documents by querying relational data. For this purpose, SQL/XML proposes three different parts. The first part provides a set of functions for mapping the data of (object-) relational tables to an XML document. The second part specifies an XML data type and appropriate functions in SQL for storing XML documents or fragments of them within a (object-) relational model. The third part describes mapping strategies of SQL data types to XML Schema data types. SQL/XML supports the requirement for specifying the representation and content of the result set but on its own (based on its alliance to SQL) there is no means for supporting multimedia retrieval in combination with MPEG-7 descriptions.

The authors in [14] propose an XML query language with multimedia query constructs called MMDOC-QL. MMDOC-QL bases on a logical formalism path predicate calculus [13] which supports multimedia content retrieval based on described spatial, temporal and visual data types and relationships. The query language defines four main clauses: OPERATION (e.g.: generate, insert, etc.) which is used to describe logic conclusions. The PATTERN clause describes domain constraints (e.g., address, etc.). Finally, there exist a FROM and CONTEXT clause which are paired together and can occur multiple times. The FROM clause specifies the MPEG-7 document and the CONTEXT is used to describe logic assertions about MPEG-7 descriptions in path predicate calculus. Of all presented query languages, MMDOC-QL fulfills best the presented requirements. Nevertheless, there are several drawbacks such as simultaneous searches in multiple databases or the integration of user preferences and usage history which are not considered in MMDOC-QL.

XIRQL [8] is a query language for information retrieval in XML documents and relies on XQL [21]. The query language integrates new features that are missing in XQL such as weighting and ranking, relevance-oriented search, data types and vague predicates and semantic relativism. A similar weighting and relevance approach has been introduced in [25].

The authors in [33] introduced *PTDOM* as a native schema aware XML database system for MPEG-7 media descriptions. Their system provides an MPEG-7 schema compliant schema catalog whose main goal is besides document validation, an appropriate typed representation of document content (elements and attributes) supporting enhanced indexing and query optimization of non-textual document content. As this paper concentrates on new improved ways for storing MPEG-7 documents, the multimedia retrieval component releis on XPath expressions and their limitations.

---

[1] http://www.sqlx.org

Besides, there exist several query languages explicitly for multimedia data such as SQL/MM [18], $POQL^{MM}$ [11] etc. which are out of scope of this chapter based on their limitation in handling XML data.

In addition to the presented MPEG-7/XML query languages, current research also concentrates on semantic multimedia retrieval and annotation. Although, MPEG-7 is probably the richest multimedia metadata set, it lacks on expressing semantic information [10, 9, 29]. As mentioned earlier, MPEG-7 relies on XML schema which is mainly used for providing structure for documents and does not impose any interpretation of the data contained [10] and therefore does not express the meaning of the structure. According to this, the authors in [29] identified open issues related to semantic multimedia annotation and introduced a three layer abstraction of the problem, indicating as top most level the use of semantic web technology (e.g., RDF, OWL, etc.).

Therefore, many authors presented approaches in order to move the MPEG-7 description of multimedia data closer to those ontology related languages [28, 30]. Related to this issue, the authors in [27] presented a user preference model for supporting semantic information.

5 MPEG-7 Query Format

MPEGs requirement paper (N2819) specified three normative parts of an MPEG-7 Query Format, namely an *Input Query Format*, an *Output Query Format* and some *Query Management Tools*. In the following, all components according to the submitted proposal of our group (University Passau, T-Systems, Fraunhofer IDMT) to MPEG will be introduced in the respective subsections. The *Query Management Tools* and its architecture are detailed in subsection 5.1. The Input and Output Query Format are combined to an MPEG-7 query language which is introduced in subsection 5.2.

5.1 MP7QF Architecture

This subsection presents an overview of all components of an MP7QF architecture. In addition to already defined parts in [3], this chapter introduces several components that have been identified as important during the evaluation of the ongoing standardization process. In this context, Figure 2 shows all elements which will be summarized in the following:

- **Session Management:** The session management provides means for the storage of user information such as user preferences and user history. In addition, the session management stores the query result for allowing relevance feedback (requirement 4.4.3 of N8219) and the search within the result set of previous searches (requirement 4.4.4 of N8219). For this purpose, the MP7QF proposal introduces a *SessionType* keeping track of those features.

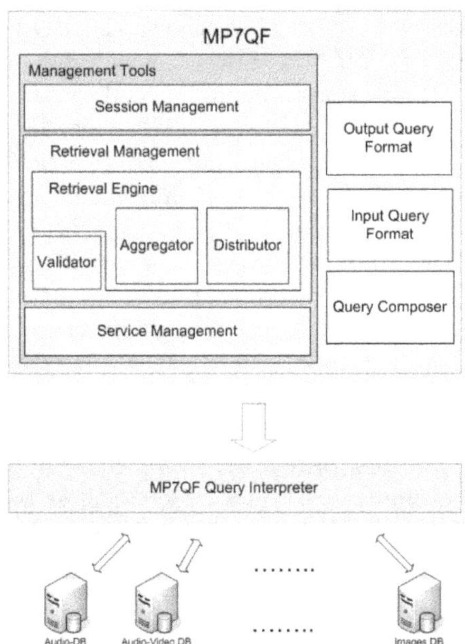

Fig. 2. MPEG-7 Query Format Framework Architecture

- **Service Management:** This component manages two parts. First, it provides methods for connecting on MP7QF-Interpreter to active user sessions. Second, the service management gathers for all active MP7QF-Interpreters their *ServiceCapabilityDescriptors*. For this purpose, every MP7QF-Interpreter must provide such a possibility, e.g., *ServiceDescriptor getServiceDescription ()*.

- **Retrieval Management:** The retrieval management component provides a collection of management tools for query validation, query aggregation and query distribution. In general, this component, besides the query validation tool, is used during the retrieval process which is instantiated by calling the search method of the retrieval engine.

- **Validator:** The validator tool allows, as implied by its name, the validation of MP7QF query descriptions which are input parameters of the retrieval process. In general, this tool is used during the query process where the conformance of the MP7QF query description is verified but may also be consulted during the creation process of the MP7QF query description.

- **Aggregator:** The aggregator is used for combining the result sets of different databases.

- **Distributor:** The distributor splits the user request into calls specific to a certain database. Note: The aggregator and the distributor are only necessary when a combined retrieval over more than one database is desired.

- **Input Query Format:** The Input Query Format (IQF) specifies the syntax and structure of an MP7 query and is detailed in subsection 5.2.
- **Output Query Format:** The Output Query Format (OQF) specifies the syntax and structure of the MP7 query result set and is detailed in subsection 5.2.
- **Query Composer:** The Query Composer defines an overall syntax and structure which combines IQF elements to one request. This request is described and transmitted with the *QueryType* to the respective MP7QF-Interpreter. The Query Composer is also used to assemble the OQF response. In addition, relevance feedback is supported by providing mechanism to integrate query results (given as OQF elements) to a new request.
- **Query Interpreter:** The MP7QF Interpreter is responsible for transforming a MP7QF query into specific bindings (e.g., XQuery, SQL, etc.) of the target databases. This necessitates that every MPEG-7 database provides such an interpreter for their query language.

5.2 MP7QF Query Language

The complete query language is specified as XML Schema and integrated within the MPEG-7 standard. Code 1 presents the entry point in form of the XML *QueryType* of an MP7QF query.

The type is mainly divided in two parts, one describing the input query format by the *input* element (of *QueryInputType*) and the other describing the output query type by the element *output* (of *QueryOutputType*).

Code 1 MPEG-7 Query Format QueryType

```
<complexType name="QueryType">
  <sequence>
    <choice>
      <element name="Input" type="m7qf:QueryInputType" />
      <element name="Output" type="m7qf:QueryOutputType" />
    </choice>
  </sequence>
  <attribute name="timeout" type="integer" use="optional"/>
</complexType>
```

It is important to note that the MP7QF *QueryType* is applied for formulating a query as well as for providing types in order to create and transfer the result set. Depending on the context (request or response) the optional child elements shall be instantiated. Further note, that information about databases that are affected by the query is provided by the *SessionType* described in [3].

Code 2 MP7QF Input Query Format in EBNF

```
Input                ::=RsPresentation,QueryCondition,SortBy;
QueryCondition       ::= {MediaResource},{Feature},
                         QueryExpression;
QueryExpression      ::= SingleSearch | CombinedSearch;
SingleSearch         ::= Operation;
CombinedSearch       ::= Operator;
Operation            ::= BrowsingQuery | TextQuery |
                         QueryByExampleMedia |
                         QueryByFeatureRange | SpatialQuery |
                         TemporalQuery |
                         QueryByFeatureDescription;
Operator             ::= BooleanOperator | CompareOperator |
                         ArithmeticOperator;
BooleanOperator      ::= "AND",Search,Search,{Search} |
                         "OR",Search,Search,{Search} |
                         "NOT",Search |
                         "XOR",Search,Search;
Search               ::= SingleSearch | CombinedSearch;
CompareOperator      ::= ("EQUAL_TO" | "GREATER_THAN" |
                         "LOWER_THAN"),
                         ({QueryByFeatureDescription},
                         ConditionOperand | ConditionValue) |
                         ArithmeticOperator;
ArithmeticOperator ::= ("SUM" | "DIVIDE" | "MULTIPLY" |
                         "SUBSTRACT"),
                         {QueryByFeatureDescription},
                         ConditionOperand | ConditionValue;
```

Input Query Format

The Input Query Format (specified by the *QueryInputType*) is responsible for describing a query request that enables user agent to query multiple MPEG-7 databases in a content-based manner. For this purpose, the query language defines three main elements, namely the *RsPresentation* element, the *QueryCondition* element and the *SortBy* element.

Result Set Presentation

The main purpose of the introduced *RsPresentationType* is to provide means for a user agent to specify the content as well as the structure of the result set. Therefore, the type supports on the one side the possibility to include XSLT stylesheets (by value or by reference) in a search request specifying how the result should be transformed and displayed. For this purpose, the *RsPresentationTransform* element of type *RsPresentationTransformType* is provided. On the other side the *RsPresentationType* provides means for identifying MPEG-7 features (descriptor and description scheme), the client wishes to receive. For this purpose, the *ReturnedMetadataType* has been introduced. The *ReturnedMetadataType* is used by a client user agent to indicate a retrieval service, which MPEG-7 feature descriptions he wishes to receive as part of the response and which structure the MPEG-7 descriptions (descriptor and description schemes) should follow. MPEG-7 feature descriptions are identified by their URN specified in the corresponding Feature Classification Scheme. The desired structure is expressed by XPath expressions.

Furthermore, the *RsPresentationType* type provides attributes to signal a client user agent's preference about the maximum number of result set entries per page (by *maxPageEntries* attribute) as well as the maximum number of result set entries overall (by *maxItemCount* attribute). The information what kind of MPEG-7 descriptors and description schemes or other metadata and transformation possibilities are supported by a specific MPEG-7 database is described in the respective *ServiceDescriptionType* introduced in [3]. The presented information is also used during query validation, checking whether the given query can be executed by the target MPEG-7 database.

Query Condition

For a better overview of most introduced XML types, Code 2 presents the MP7QF Input Query Format in EBNF notation whereas it has to be noted that this is only an extraction of the complete MP7QF Schema. As already mentioned in previous sections, the entry point for a query is the *QueryType* consisting of input and output elements. This paragraph concentrates on the input element and here especially on the query condition. In general, a query condition may consist of a set of media resources, a set of features and exactly one query expression. A media resource represents example media files (e.g., an image or audio file) that are passed by the client application.

A feature represents one MPEG-7 descriptor or description scheme. Finally, the query expression defines the syntax and semantic of the retrieval criteria. As indicated by the EBNF, the query expression consists of either a single search or a combined search. A single search is the most primitive construct and instantiates one of the provided query operations. A detailed explanation of some selective query operations and their use can be found in the following paragraphs.

The second possibility is a combined search. A combined search reflects the combination of an operator with one or more ancillary single or combined searches. For extensibility purpose, the MP7QF Schema provides a mechanism for query operations that extend the abstract *OperationBaseType* as well as for query operators that extend the abstract *OperartorBaseType*.

Query Types

Our submitted proposal covered a large set of different query types such as *TextQueryType*, *SpatialQueryType*, *TemporalQueryType*, *QueryByFeatureRangeType*, etc. In order to deliver insight, three of them are introduced in more detail.

- **BrowsingQueryType**
  The *BrowsingQueryType* extends the abstract *OperationBaseType* and
  can substitute a *SingleSearch* within a query expression. This type realizes browsing (empty query) of the multimedia content and requests the
  database for a selection of available multimedia data. A client user agent
  has the possibility to restrict the selection to a specific domain by using
  the *media* attribute. This attribute can use one of the following values,
  namely *textual, image, audio, video, audiovisual* or *any*. The value *any*
  denotes that the user is interested on a selection of multimedia data in
  any domain.
- **QueryByExampleMediaType**
  The *QueryByExampleMediaType* extends the abstract *OperationBaseType*
  and can substitute a *SingleSearch* within a query expression. The *Query-
  ByExampleMediaType* enables a user agent to realize similarity search
  based on a given multimedia datum. The example multimedia data must
  be described by the *MediaResource* element and referenced within the
  QueryByExampleMediaType by the *mediaIDREF* attribute. Note, that the
  target database is responsible for extracting necessary low- and/or high-
  level descriptors for similarity matching against stored data. The type of
  operation (exact match or similarity) is indicated by a Boolean attribute
  named *exactMatch*.
- **QueryByFeatureDescriptionType**
  The *QueryByFeatureDescriptionType* extends the abstract *OperationBase-
  Type* and can substitute a *SingleSearch* within a query expression. The
  QueryByFeatureDescriptionType enables a user agent to realize similarity
  retrieval based on a given MPEG-7 description (descriptor or description
  scheme). The example MPEG-7 description must be described by the *FeatureType* element and referenced within the *QueryByFeatureDescription-
  Type* by the *featureIDREF* attribute. The second attribute (*exactMatch*)
  denotes whether the target database should trigger an exact match operation or not. However, the algorithm used for similarity matching is
  incumbent on the respective database system and can vary from instance
  to instance.

Query Operators

- **Boolean Operators**
  The *BooleanOperatorType* extends the abstract *OperatorBaseType* and
  provides means for the combination of *SingleSearch* elements and *CombinedSearch* elements according the given Boolean operators AND, OR,
  NOT and XOR. Note, that the Boolean operators AND and OR have a
  minimum of 2 elements and an unbounded maximum. Whereas the XOR
  Boolean operator is fixed to exactly two single or combined searches and
  the NOT Boolean operator is limited to exactly one element.
- **Comparison Operators**

Similar to the *BooleanOperatorType*, the *ComparisonOperatorType* extends the abstract *OperatorBaseType* and provides means for comparing elements of MPEG-7 descriptors and description schemes according to a given value. It allows for instance queries like: Give me all audio descriptions whose AudioBPMType has a scalar value greater than 120.

When we have a deeper look to a comparison expression e.g., $A > B$ within our MP7 query language, we can identify 3 different meanings for an operand A. First of all, the operand can specify a specific element of a given MPEG-7 descriptor or descriptor scheme (indicated by the Feature element). For this purpose, the comparison operator would be used in combination with the *QueryByFeatureDescription* element referencing to a *FeatureType* holding the MPEG-7 descriptor and the *ConditionOperand* pointing by a XPath expression to the desired MPEG-7 element within the given MPEG-7 descriptor.

Second, the operand can represent a single alphanumeric value (e.g., 120) either by using the *ConditionValue* element or by using an *ArithmeticOperator* within a *CombinedSearch* element.

Third, the operand can reference to an element (hence to its value) of an MPEG-7 descriptor that is already available at the underlying database schema. In this case, one uses only the *ConditionOperand* pointing by a XPath expression to the desired MPEG-7 element

Note that this operator type is restricted to *SingleSearch* elements of type *QueryByFeatureDescriptionType*.

- **Arithmetic Operators**
  The *ArithmeticOperatorType* extends the abstract *OperatorBaseType* and provides means for calculating alphanumeric values that can be used as comparison operands. Supported operators are *SUM*, *DIVIDE*, *SUBTRACT* and *MULTIPLY*. Similar to a comparison expression, an arithmetic expression, e.g., $A * B$ supports several fields of application. For instance, by using the *QueryByFeatureDescription* element in combination with the *ConditionOperand*, one can reference to a specific value of an MPEG-7 descriptor. In addition, one has the possibility to use the *ConditionOperand* alone, highlighting to use the corresponding value in the database. Furhtermore, the *ConditionValue* element is available for representing single alphanumeric values.

Output Query Format

The OQF is defined by the *QueryOutputType*. It provides a container for all response related information and comprises media resources (by reference or by value) and result set items or groups thereof. Media resources are referenced from within result set items to enable a clean separation of media data and describing metadata. Elements of type *RsGroups* can be used to group single items of the result set e.g. to support paging of the result set.

RsGroupType

The *RsGroupType* provides means for grouping one ore more result set items (e.g. that appear on the same result set page). Alternatively it encapsulates a transformed result page that is the result of an XSLT transformation applied to a result page in order to generate an output format customized for a specific client user agent. In addition, the *RsGroupType* provides two attributes for supporting paging of the result set. The first, *totalPages* indicates the maximum amount of pages the server has available. The second, *curPage* identifies a single *RsGroup* within the available set of pages.

RsItemType

The *RsItemType* represents a single item within the result set. It encapsulates optionally MPEG-7 feature descriptions as *DescriptionUnits* or other metadata, that is not specified by MPEG. Each *RsItem* has a required attribute *rank* indicating the overall rank of this item within the result set. Optionally it provides the attribute *confidence* indicating the services confidence value for the match, *rsrefURI* for referencing a media resource on the service side (instead of including it by value in the result set structure) and *rsrefID* for referencing a media resource that is part of the structured result set and available in a *MediaResource* element of the OQF instance.

5.3 Examples

This subsection demonstrates the use of the MP7 Query Format based on scenarios and use cases common in multimedia retrieval systems. The imaginary database used for retrieval supports any kind of MPEG-7 descriptor and descriptor scheme. In addition, all proposed MP7QF Query Input operations and operators are permitted. All subsequent examples base on the described database scenario meaning that there is no restriction in any direction (MPEG-7 types, IQF operations or operators).

RsPresentation

This example in Code 3 presents the use of *RsPresentationType* in order to enable a user to specify the content as well as the structure of the query result. The given example demonstrates the use of the *MPEG7Feature* element determining the content of the result set by indicating the corresponding classification scheme element (e.g., urn:mpeg:mp7qf:2003:3:1). In this example, it means that the result set should contain MPEG-7 *Image* information. In addition, we are estimating the *CreationInformation* of annotated *Images* that satisfy a given condition. Note, the condition is not shown in this example.

Code 3 RSPresentation example

```
<m7qf:Query timeout="200">
  <m7qf:Input>
    <m7qf:RsPresentation >
        <m7qf:ReturnedMetadata>
    <m7qf:MPEG7Feature csRef="urn:mpeg:mp7qf:2007:3.1">
                    ./child::Image
    </m7qf:MPEG7Feature>
    <m7qf:MPEG7Feature csRef="urn:mpeg:mp7qf:2007:5.4">
                    ./child::Image/child::CreationInformation
    </m7qf:MPEG7Feature>
        </m7qf:ReturnedMetadata>
      </m7qf:RsPresentation>
      ()
  </m7qf:Input>
</m7qf:Query>
```

Query By Example Descriptor Scenario

This example query in Code 4) demonstrates a query by example request whose input is an MPEG-7 descriptor (*DominantColorType*). The MPEG-7 descriptor is instantiated by the *Feature* element and referenced within the specific query operation (in this example a *QueryByFeatureDescriptionType*) by the *featureIDREF* attribute. Note that in this example the *exactMatch* attribute is set to false indicating that a similarity search based on the given dominant color should be performed. Further, note that the query language does not define which similarity algorithm should be used.

Result Set Secnario

The previous examples presented the formulation of a client request to one or multiple MPEG-7 database(s). The following examples demonstrate how an MPEG-7 database can respond to such a request and how a result set can be transmitted in a standardized way. As described in subsection 5.2, the query language provides for this purpose the *QueryOutputType* and its elements of type *RsGroupType* and *RsItemType*. Note, that the global *QueryType* is used as container for the client request as well as for the database response.

The example in Code 5 demonstrates a possible result set containing MPEG-7 descriptions with additional group structuring by the *RsGroup* element. The *RsGroup* element is optional and a possible result set also can contain only *RsItem* elements holding MPEG-7 descriptions. The *RsGroup* element provides two main advantages. First, by using the attributes *curPage* and *totalPages* paging of the result set is supported and second (not shown in this example), by using the *TransformedResultPage* element, the server can assign specific XSLT stylesheets for transformation or already provide the result set in a transformed manner e.g. as HTML representation.

Code 4 Query By Example Descriptor Example

```
<m7qf:Query timeout="200">
 <m7qf:Input>
   <m7qf:RsPresentation maxPageEntries="5">
     ()
   </m7qf:RsPresentation>
   <QueryCondition>
     <m7qf:Feature id="dc1">
       <m7qf:VisualD xsi:type="DominantColorType">
         <ColorSpace type="RGB"/>
         <SpatialCoherency>28</SpatialCoherency>
         <Value>
           <Percentage>12</Percentage>
           <Index>1 1 1</Index>
           <ColorVariance>1 1 1</ColorVariance>
         </Value>
       </m7qf:VisualD>
     </m7qf:Feature>
     <m7qf:QueryExpression>
       <m7qf:SingleSearch xsi:type="m7qf:QueryByFeatureDescriptionType"
           featureIDREF="dc1" exactMatch="false" />
     </m7qf:QueryExpression>
   </QueryCondition>
 </m7qf:Input>
</m7qf:Query>
```

Code 5 Result Set Example

```
<m7qf:Query>
   <m7qf:Output>
       <m7qf:RsGroup currPage="1" totalPages="1">
           <m7qf:RsItem rank="1"  >
               <m7qf:DescriptionUnit xsi:type="ImageType">
                   <Image>
                       ()
                   </Image>
               </m7qf:DescriptionUnit>
           </m7qf:RsItem>
       </m7qf:RsGroup>
   </m7qf:Output>
</m7qf:Query>
```

6 Current Status of the MP7QF Committee Draft

The standard document has already reached the status CD of the MPEG standardization committee. Therefore the core of the framework can be considered as stable. In the following, the main concepts are described and their relation to our proposal is highlighted.

6.1 Query Format

Similar as MPEG-7, the MP7 query format is fully compatible to XML schemas. The query format is destined to transmit MPEG-7 descriptors and description schemes. Nevertheless, a mechanism has been established to allow transmitting information from other schemas than MPEG-7 as well. The root node of the query format schema consists of an input type and an output type. The input format describes the query interface from the user to the database (see section 6.2), whilst the output query format (see section 6.3) describes the response from the server to the client. A further attribute has been implemented in order to specify an ID value for the query. This number could be used e.g. in an asynchronous mode, where the server not directly responds to the client and closes the connection. After a certain time, the client could ask again for the result with the previous ID.

6.2 Input Query Format

The input query format has been established in order to enable the communication between client and server. In this part of the query formulation, different operations and operators are handled, descriptions and content are defined and the expected result set is transmitted. The input query format node is named *InputType* and consists of three types: *QFDeclarationType*, *OutputDescriptionType* and *QueryConditionType*. These three types have been placed into a sequence in order to allow transferring a number of queries at the same time. *QFDeclarationType* (similar to our *MediaResource* and *Feature* types) allows to reference the actual content (e.g. MPEG-7 descriptions or audiovisual content). They have been established in order to describe the content once, and in series can be used within the actual query declaration as reference a couple of times without writing the same descriptor every time. The proposed type consists of a sequence of the type *ResourceType* and an attribute with an id as referencing number. The *ResourceType* can contain a choice of an MPEG-7 descriptor, a *TextAnnotation*, a *MediaResource* or an *AnyDescription*, so every kind or desired resource is supported for referencing. Especially to mention is the *AnyDescription* type. This type enables the use of own descriptors of arbitrary XML schemas.

The *OutputDescriptionType* is the definition of the expected output (similar to our *RsPresentation* type) in order to give the user the opportunity to specify the format of the output he expects, when submitting the query.

The expected output can be a resource locator, freetext or metadata from any namespace. The *QueryConditionType* provides means for expressing restrictions and conditions in order to shrink the result set to the user's needs. The node consists of a sequence of *ConditionType* elements, which is an abstract base class. Inherited from *ConditionType* are the classes *OperatorType* (see our abstract *OperatorBaseType*), which describes the operators link the actions and *OperationType* (see our *OperationBaseType*) that describes the actual condition. An example for this kind of type is the *QueryByExampleType*, which looks for similar or exact items of the qiven example. At the moment five different types have been implemented into the CD. Besides the *QueryByExampleType*, a *QueryByFreeTextType* has been defined in order to allow a freetext search. Furthermore an *XQueryType* has been implemented to search with XQuery [31] expressions. The last defined type is the *QueryByRelevanceFeedbackType*, which describes a query operation that takes into consideration the result of the previous retrieval. The *OperatorType* is an abstract base type and serves as root type for all operator classes such as Boolean, Comparison or Arithmetic types. Currently, only Boolean operators have been defined. Among these are AND, OR, NOT and XOR operators (similar to our Boolean operator types).

6.3 Output Query Format

The OQF specifies the expected output from the server to the client and the node within the XML schema is called *OutputType*. This node consists of a sequence of three different elements, named *GlobalFreeText*, *SystemMessage* and *ResultItem*. The *GlobalFreeText* element consists a string and specifies a text message that the server may want to reply to the client. The *SystemMessage* element consists of the type *SystemMessageType*, which replies a message to the client. This can be a choice of status messages, error messages and warnings. The *ResultItemType* specifies the actual result and consists of a sequence of *Resource*, *FreeText* and *Description* elements. The *Resource* element contains an URI of the object, the *FreeText* element contains a textual description and the *Description* element of namespace "'any"' in order to be able to reply descriptors from other schemas as MPEG-7 as well.

6.4 Query Examples

The example presented in Code 6 illustrates some key aspects of the current MP7QF CD. The namespaces declaration of the *Mpeg7Query* element has been removed in order to increase readability. The presented query specifies the need to find similar multimedia items to the given MPEG-7 *DominantColorType* descriptor. Note, the query is not restricted to a specific multimedia type (e.g., audio) and therefore, could result in images with a similar color representation as well as in audio annotations whose CD cover image has a similar representation. Instead of inlining the example within the *Condition*

Code 6 Example usage of the *QueryByExample* QueryType

```
<Mpeg7Query>
 <Input>
  <QFDeclaration>
   <Resource id="id1">
    <MPEG7DescriptionElement
                  xsi:type="mpeg7:DominantColorType">
      <mpeg7:ColorSpace type="RGB"/>
      <mpeg7:SpatialCoherency>28</mpeg7:SpatialCoherency>
      <mpeg7:Value>
        <mpeg7:Percentage>12</mpeg7:Percentage>
        <mpeg7:Index>1 1 1</mpeg7:Index>
        <mpeg7:ColorVariance>1 1 1</mpeg7:ColorVariance>
      </mpeg7:Value>
    </MPEG7DescriptionElement>
   </Resource>
  </QFDeclaration>
  <OutputDescription
     outputNameSpace="urn:mpeg:mpeg7:schema:2004">
   <Field typeName="CreationType">/Title</Field>
  </OutputDescription>
  <QueryCondition>
   <Condition xsi:type="QueryByExample">
     <ResouceREF>id1</ResouceREF>
   </Condition>
  </QueryCondition>
 </Input>
</Mpeg7Query>
```

element (which is also possible), it is defined within the *QFDeclaration* section. This would allow, for example, to reuse the same data for other conditions just by referring to the same *id*. The desired output (content as well as structure) is denoted by the *OutputDescription* element. In this example, the query specifies the field *Title* from the MPEG-7 schema, which should appear in the result items if available.

7 Conclusion

This chapter introduced an MPEG-7 query format (MP7QF) architecture and query language based on the defined requirements at the 77th MPEG meeting in July 2006. The query language provides means for interoperability among distributed multimedia search and retrieval services and makes intelligent content navigation in MPEG-7 enabled multimedia repositories possible that are loosely connected via distributed heterogenous networks. In addition, parts of our MP7QF XML Schema have been introduced such as *QueryInputType*, *QueryOutputType* or *QueryExpressionType* and the main concepts and scenarios were elaborated. More detailed information about this work is expected in the near future as the work is at present progressed within the MPEG standardization. Results of this collaborative process are expected to be available by mid 2008 with the finalization of that new proposed ISO/IEC standard (part 12 of MPEG-7 standard).

References

1. James Clark and Steve DeRose. XML Path Language (XPath). *W3C Recommendation, http://www.w3.org/TR/xpath*, 1999.
2. Alin Deutsch, Mary Fernandez, Daniela Florescu, Alon Levy, and Dan Suciu. XML-QL: A Query Language for XML. *W3C, http://www.w3.org/TR/1998/NOTE-xml-ql-19980819/*, August 1998.
3. Mario Döller, Ingo Wolf, Matthias Gruhne, and Harald Kosch. Towards an MPEG-7 Query Language. In *Proceedings of the International Conference on Signal-Image Technology and InternetBased Systems (IEEE/ACM SITIS'2006)*, pages 36–45, Hammamet, Tunesia, 2006.
4. Andrew Eisenberg and Jim Melton. SQL/XML is Making Good Progress. *ACM SIGMOD Record*, 31(2):101–108, June 2002.
5. N. Fatemi, O. Abou Khaled, and G. Coray. An XQuery Adaptation for MPEG-7 Documents Retrieval. *XML Conference & Exposition 2003, Philadelphia, PA, USA*, December 2003.
6. Nastaran Fatemi, Mounia Lalmas, and Thomas Rlleke. How to retrieve multimedia documents described by MPEG-7. In *Proceedings of the 2nd ACM SIGIR Semantic Web and Information Retrieval Workshop, ACM Press,,* New York, NY, USA, 2004.
7. Nastarn Fatemi, Omar Abou Khaled, and Giovanni Coray. An XQuery Adaptation for MPEG-7 Documents Retrieval. In *Proceedings of the XML Conference and Exposition*, Philadelphia, PA, USA, 2003.
8. Norbert Furh and Kai Grossjohann. XIRQL: A Query Language for Information Retrieval in XML Documents. In *Proceedings of the 24th ACM-SIGIR Conference on Research and Development in Information Retrieval*, pages 172–180, New Orleans, Louisiana, USA, 2001.
9. Roberto Garcia and Oscar Celma. Semantic Integration and Retrieval of Multimedia Metadata. In *Proccedings of the 5th International Workshop on Knowledge Markup and Semantic Annotation (SemAnnot 2005)*, Galway, Ireland, November 2005. CEUR Workshop Proceedings.
10. Samira Hammiche, Salima Benbernou, Mohand-Said Hacid, and Athena Vakali. Semantic Retrieval of Multimedia Data. In *Proceedings of the second ACM International Workshop on Multimedia Databases, ACM-MMDB 2004*, pages 36–44. ACM Press, 2004.
11. Andreas Henrich and Gnter Robbert. *POQL^{MM}*: A Query Language for Structured Multimedia Documents. In *Proceedings 1st International Workshop on Multimedia Data and Document Engineering (MDDE'01)*, pages 17–26, July 2001.

12. Harald Kosch and Mario Döller. The MPEG-7 Multimedia Database System (MPEG-7 MMDB). *Journal of Systems and Software, Accepted for publication, In Press by Elsevier. To appear in spring 2007*, 2007.
13. Peiya Lui, Amit Charkraborty, and Liang H. Hsu. Path Predicate Calculus: Towards a Logic Formalism for Multimedia XML Query Language. In *Proceedings of the Extreme Markup Languages*, Montreal, Canada, 2000.
14. Peiya Lui, Amit Charkraborty, and Liang H. Hsu. A Logic Approach for MPEG-7 XML Document Queries. In *Proceedings of the Extreme Markup Languages*, Montreal, Canada, 2001.
15. Matthias Lux, Werner Klieber, and Michael Granitzer. Caliph & Emir: Semantics in Multimedia Retrieval and Annotation. In *Proceedings of the 19th International CODATA Conference 2004: The Information Society: New Horizons for Science* , pages 64–75, Berlin, Germany, 2004.
16. J. M. Martinez. *MPEG-7 Overview*. ISO/IEC JTC1/SC29/W11 N5525, Pattaya, March 2003.
17. J. M. Martinez, R. Koenen, and F. Pereira. MPEG-7. *IEEE Multimedia*, 9(2):78–87, April-June 2002.
18. Jim Melton and Andrew Eisenberg. SQL Multimedia Application packages (SQL/MM). *ACM SIGMOD Record*, 30(4):97–102, December 2001.
19. Ravi Murthy and Sandeepan Banerjee. XML Schemas in Oracle XML DB. In *Proceedings of the 29th VLDB Conference*, pages 1009–1018, Berlin, Germany, 2003. Morgan Kaufmann.
20. Oracle. Oracle Database 10G Release 2, XML DB. *http://download-uk. oracle. com/otndocs/tech/xml/xmldb/TWP_XML_DB_10gR2_long.pdf*, 2006.
21. Jonathan Robie. XQL (XML Query Language). *http://www. ibiblio. org/ xql/xql-proposal. html*, 1999.
22. Albrecht Schmidt, Martin Kersten, Menzo Windhouwer, and Florian Waas. Efficient relational storage and retrieval of XML documents. *Lecture Notes in Computer Science*, 1997:137+, 2001.
23. Harald Schning. Tamino - a DBMS designed for XML. In *Proceedings of the 17th International Conference on Data Engineering (ICDE)*, pages 149–154, April 2001.
24. K. Staken. Xindice Developers Guide 0.7. *The Apache Foundation, http: //www. apache. org*, December 2002.
25. Anja Theobald and Gerhard Weikum. Adding Relevance to XML. In *Proceedings of the 3rd International Workshop on the Web and Databases (WebDB)*, pages 35–40, Dallas, USA, 2000.
26. Belle L. Tseng, Ching-Yung Lin, , and John R. Smith. Video Personalization and Summarization System. In *Proceedings of the SPIE Photonics East 2002 - Internet Multimedia Management Systems*, Boston, USA, 2002.
27. Chrisa Tsinaraki and Stavros Christodoulakis. A Multimedia User Preference Model that Supports Semantics and its Application to MPEG 7/21. In *Proceedings of the 1st International Workshop on Semantic Media Adaptation and Personalization*, Athens, Greece, December 2006.
28. Chrisa Tsinaraki, Panagiotis Polydoro, Nektarios Moumoutzis, and Stavros Christodoulakis. Coupling OWL with MPEG-7 and TV-Anytime for Domain-specific Multimedia Information Integration and Retrieval. In *Proceedings of RIAO 2004*, Avignon, France, April 2004.

29. Jacco van Ossenbruggen, Giorgos Stamou, and Jeff Z. Pan. Multimedia Annotations and the Semantic Web. In *Proccedings of the ISWC Workshop on Semantic Web Case Studies and Best Practices for eBusiness, SWCASE05*, Galway, Ireland, 2005.
30. Shankar Vembu, M. Kiesel, Michael Sintek, and Stephan Baumann. Towards Bridging the Semantic Gap in Multimedia Annotation and Retrieval. In *Proceedings of the 1st International Workshop on Semantic Web Annotations for Multimedia, SWAMM 2006*, Edinburgh, Scotland, May 2006.
31. W3C. XML Query (XQuery). *W3C, http://www.w3.org/TR/xquery/*, 2006.
32. Utz Westermann and Wolfgang Klas. An Analysis of XML Database Solutions for the Management of MPEG-7 Media Descriptions. *ACM Computing Surveys*, 35(4):331–373, December 2003.
33. Utz Westermann and Wolfgang Klas. PTDOM: a schema-aware XML database system for MPEG-7 media descriptions. *Software: Practice and Experience,*, 36(8):785–834, 2006.
34. Ling Xue, Chao Li, Yu Wu, and Zhang Xiong. VeXQuery: an XQuery extension for MPEG-7 vector-based feature query . In *Proceedings of the International Conference on Signal-Image Technology and InternetBased Systems (IEEE/ACM SITIS'2006)*, pages 176–185, Hammamet, Tunesia, 2006.

Part IV

Cross-Modal Multimedia Techniques

Visualisation Techniques for Analysis and Exploration of Multimedia Data

Vedran Sabol, Wolfgang Kienreich, Michael Granitzer

Know-Center Graz, Inffeldgasse 21a, 8010 Graz, Austria
{vsabol,wkien,mgrani}@know-center.at

Information technology has developed past its traditional focus on text-based data. The much-cited rapid growth of available information has been accompanied by a diversification of information types. Multimedia data is rapidly becoming the predominant form of information created, processed and distributed in many application domains.Multimedia data sets are characterized by their heterogeneous nature and complex structure. Documents often combine of different modalities, for example video streams, audio streams and textual information. Document content often features a pronounced temporal component, as in the case of audio and video data. Multimedia documents frequently include rich semantic descriptors and complex structures of cross-modal, inter- and intra-document references.

It is often not feasible to manually annotate such complex semantics, especially in the context of very large data sets. Various automated methods for the extraction of semantic metadata have been proposed and evaluated. However, the capabilities of automatic extraction are limited in terms of accuracy, performance and diversity of results. Visualisation techniques employ the vast processing power of the human visual apparatus to quickly identify complex patterns in large amounts of data. When combined with machine processing capabilities, such techniques provide unparalleled means for gaining insight into large data sets in general, and into multimedia data sets in particular.

The multi-faceted nature of multimedia documents has led to a variety of visual representations for navigating, analysing and understanding of multimedia data sets. As each representation is specifically designed to address different aspects of the data, innovative approaches combining

V. Sabol et al.: *Visualisation Techniques for Analysis and Exploration of Multimedia Data*, Studies in Computational Intelligence (SCI) **101**, 219–238 (2008)
www.springerlink.com

several visualisations in a single coordinated interface had to be introduced. This chapter presents a comparative discussion of selected multimedia visual representations and tools.

1 Introduction

Recent studies estimate the amount of information globally available to grow from 5 Exabyte in 2003 [29] and 161 Exabyte in 2006 to 998 Exabyte until 2010 [17]. Most of the new information created, shared and archived is of a multimedia nature. For example, 33,000 worldwide TV stations produced approximately 48 million hours of broadcasted video content in 2002 [51]. This already impressive number acquires still more significance considering that a mere 2000 hours of CNN news videos consist of approximately 1.7 million shots and almost 70,000 stories [52]. Another, less obvious, example of multimedia information influx is provided by video surveillance at some 14,000 worldwide air terminals, which produces 4.5 million hours of video per day [51].

To quantify the distribution of information types, 75% of available content consisted of images and video streams and 20% was contributed by voice recordings in 2006. It is interesting to note that 95% of the available content is comprised of "unstructured data" like digital images, music tracks, voice packets and video broadcasts. Such content is not organized into indexed structures, as for example database records, and can not be uniquely identified and accessed. Fortunately, an increasing number of multimedia documents is annotated by producers and users, for example using time stamps, geographical references or semantic metadata.

Requirements for searching, browsing and displaying information in this context obviously differ from traditional, text-based systems. Research in the area has to provide innovative, multimedia-oriented methods for indexing and locating documents. In addition, new visualisation approaches and user interface technologies are required to enable users to explore, access and understand retrieved multimedia information.

The areas of information visualisation and human computer interaction are introduced in the remainder of this section, both establishing a foundation on which multimedia visualisation is built upon. The following section attempts to define multimedia visualisation and provides more details on its peculiarities as well as on manifold aspects (dimensions) of data it is dealing with. Finally, a selection of multimedia visualisation examples is presented and briefly discussed. The chapter concludes with a discussion of the current state of multimedia visualisation and gives an outlook on further development.

1.1 Information Visualisation

Visualisation is commonly defined as the use of visual representations to aid in analysis of quantitative or qualitative information. Starting with historical efforts to annotate geographical or celestial maps, various applications of this idea are documented. The advent of computer graphics greatly increased interest in the area because interactive visual representations of large amounts of information became possible. Information visualisation emerged as a new discipline in computer science addressing this development.

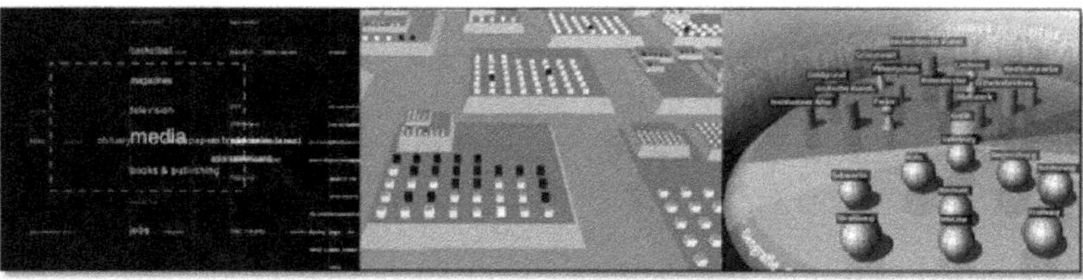

Fig. 1. Information visualisation history - examples. From left to right: Visualisation of news articles (1994) [39], visualisation of directory structures (1997) [3], Visualisation of Encyclopedia Knowledge spaces (2005) [22]

Information visualisation is commonly defined as the use of computer-supported, interactive, visual representations of abstract data, information spaces and structures to amplify cognition. The focus of information visualisation is on visual abstractions and metaphors and corresponding interaction models for exploration and analysis of abstract information and data structures. This differentiates information visualisation from scientific visualisation, which usually deals with numerical data modelling real-world phenomena (as in meteorology) or sensory data (as in medical visualisation).

Based on the observation that the human visual apparatus is capable of rapid perception, transformation and pattern recognition of information, work in the field literally tries to enable "the use of vision to think" [8]. Beside the dedication to computer-supported methods, information visualisation is distinguished by a clear focus on abstract information and on information seeking. Some key insights gained in the field are perhaps best described by the *"mantra of visual information seeking"* as stated by Ben

Shneiderman: *"Overviews first, zoom and filter, and details on demand"* [45].

Today, information visualisation as a discipline has reached a stage of development where sub-areas start to emerge which differ significantly in information base, algorithmic approach and application area. Contrasting the focus of information visualisation on large information spaces and individual information seeking, the comparably new field of knowledge visualisation focuses on the transfer of knowledge among persons [7]. Knowledge visualisation often works on smaller, but highly organized sets of information and employs multiple complementary visualisations which convey different aspects of a common knowledge space. Commonly understood as the legitimate successor to classical information visualisation, the emerging field of visual analytics is concerned with the use of abstract visual metaphors in combination with a human information discourse to enable detection of patterns in massive, dynamically changing information spaces [47]. In this context sense-making and reasoning play a prominent role. To a large part, visual analytics is powered by the increasing integration of data mining technology and information visualisation methods.

1.2 Human-Computer Interaction

Human-computer interaction (HCI) is commonly defined as "a discipline concerned with the design, evaluation and implementation of interactive computing systems for human use and with the study of major phenomena surrounding them" [16]. It is an interdisciplinary research area centred around computer science in general and computer graphics in particular. Yet, HCI is influenced by a number of other fields such as psychology, cognitive and perceptual sciences as well as design. The primary goal in Human-computer interaction is to find ways to overcome the barriers between humans and computers by simplifying the way in which humans accomplish their tasks using computers. Usability, the ease of using a tool (for example a program or a Web page), to achieve a specified goal, has emerged as a key concept in Human-computer interaction. This concept has been formally defined in an ISO-Standard [19]: "Usability: the extent to which a product can be used by specified users to achieve specified goals with effectiveness, efficiency and satisfaction in a specified context of use". Usability engineering performs optimizations to design and implementation of interfaces and interaction models and evaluates the effectiveness of optimizations through a variety of methods (for example, heuristic evaluations or formal studies). Ideally, phases of design, implementation and evaluation form an iterative process which generates applications that are easily accessible to the desired target group.

2 Multimedia Visualisation

2.1 Definition

Multimedia visualisation has yet to be defined in clear and certain terms. Most experts in the field readily agree that multimedia visualisation is deeply rooted, in its methods and approaches, in disciplines like information visualisation, human-computer interaction and information design. Consensus quickly disintegrates, however, when it comes to defining of boundaries between multimedia visualisation and related areas. Historically, multimedia visualisation has often been understood as the visualisation of multimedia entities, which were equated to audio and video files. This view was formed in times when numerical and textual data was the norm, and multimedia data was the exception present only in specialized application areas like video editing.

The advent of affordable digital audio and video capture devices and the resulting avalanche of multimedia data have quickly invalidated this view on multimedia visualisation. Today, most areas in the visualisation domain have been reshaped by the prevalence of multimedia data. For example, visual analytics and knowledge visualisation, two recently formed research branches in information visualisation, both stress the need to visualise multimedia data in their respective research agenda. Consequentially, a modern definition should probably place multimedia visualisation closer to the centre and define other visualisation disciplines as sub areas specializing in certain ways.

The rise of semantic technologies has shaped an application environment for retrieval and visualisation systems in which the actual format of content becomes increasingly irrelevant. Still, the peculiarities of content formats demand customization of visualisations and invite specialized forms of visual representation. Consequentially, a modern definition should probably stress that multimedia visualisation is concerned with utilizing specific properties of media to create visual representations which aide exploration and interpretation of large, multimodal repositories. Given the prevalence of multimedia information its adoption by various branches of the visualisation field, it is likely that a coherent definition of multimedia visualisation will be formulated in the near future.

2.2 Dimensions

Multimedia documents typically feature a variety of modalities and are annotated with metadata, which is often of a rich semantic nature. As a consequence, multimedia documents expose several dimensions relevant for designing, implementing and evaluating visualisation. This section briefly discusses some of these aspects.

Content: Just like textual content, multimedia content is primarily human-readable. Its structure is mainly defined through sequences (for example words, frames or samples), through spatial and temporal constraints and through intra- and inter-document links. Multimedia content is usually loosely structured. Thus, as of today (2007), it is far from machine-understandable. Multimedia content generally requires sophisticated presentation and processing tools, and extraction of features and recognition of concepts is far more difficult than for textual content. This is further aggravated by the fact that multimedia documents often contain different modalities (for example video with audio tracks and textual annotations) with "incompatible" sets of extracted features.

Semantics: Multimedia content can be annotated by rich semantic metadata. Metadata standards, such as MPEG-7 and ontologies, are employed to formalize semantics for multimedia content description. Therefore, multimedia visualisation must be capable of handling semantic structures along the relationships and patterns arising from the actual content.

Temporal: While texts and images may feature a time-stamp or contain temporal references, audio and video are inherently "temporal" since they have duration and a well defined sequence. The visualisation of temporal data is a distinguished research area of its own [33]. Results from this area often influence work in multimedia visualisation.

Geospatial: Because multimedia content is often related to locations and places in some way (most commonly the origin of a video or a photo), geospatial metadata plays a distinguished role in multimedia visualisation. Similar to temporal data visualisation, geospatial visualisation is a fairly complex research area in its own right. This further underpins the interdisciplinary character of multimedia visualisation.

Other dimensions: As multimedia data often represents or captures real world scenarios, visualisation of abstract concepts and structures, as in information visualisation, may not be sufficient: 3D models (for example in architecture) or sensory data (such as in medicine or physics) may also be a component of a multimedia dataset. As a result domain specific scientific visualisation may also play a role in multimedia visualisation.

3 Selected Examples of Multimedia Visualisation

In this section an overview of different systems for visualisation of multimedia data is given. The systems are grouped by the task they are designed for, and for each task one or more selected examples are shortly described.

3.1 Browsing and Navigating Videos

Tools for browsing and non-linear editing of videos have become ubiquitous (for example Adobe Premiere, Final Cut or Pinnacle Studio lines). In their simplest form they are present on many desktop computers as they are incorporated into the operating system (Windows Movie Maker or iMovie). A central visual component, which is present on almost all systems for browsing or manipulation of videos, is the timeline. A timeline is a chronological representation of the progress of a video, typically displaying a sequence of thumbnails extracted from the videos at different time points. Additional information, such as audio activity may also be integrated into the timeline.

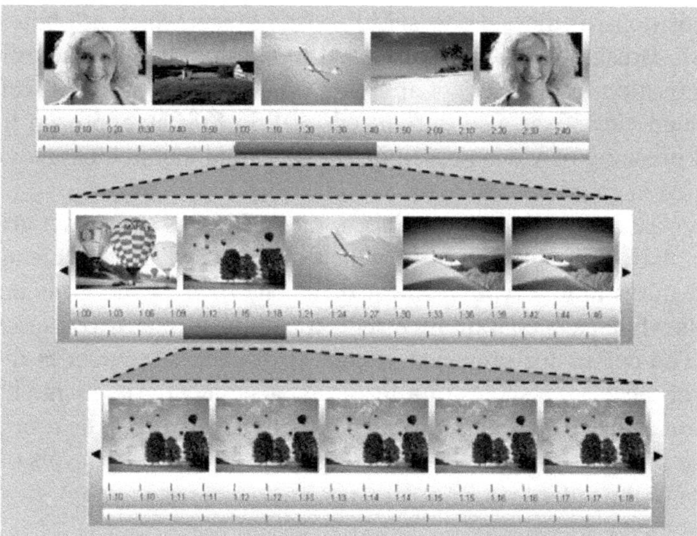

Fig. 2. Schematic of hierarchical timeline visualization

Silver Video Editor [9] offers a hierarchical timeline view (first introduced in [31]). Its principle is illustrated in Fig. 2: The top level timeline represents the video in its full duration; increasing, user-defined levels of details are displayed in the timelines below the top-level one for selected ranges.

Successor projects have extended the concept with multiple-lens semantic zooming capabilities, where selected parts of the video are magnified within a single (non-hierarchic) timeline [27].

Fig. 3. Schematic of the movieDNA visual representation

Brushing can be defined as a collection of powerful investigative techniques over user interface elements for querying and selecting a subset of objects displayed within a visualisation. The user applies the brush to different positions of the view to select those objects which are bounded by the brush. Brushing was originally used for exploration of scatterplots but a generalisation of the concept applying on other view types as well as on coordinated views was also introduced [40]. In [38] an approach for hierarchical brushing of videos based on the so called hierarchical movieDNA (illustrated in
Fig. **3**) visual abstraction is introduced. movieDNA is a compact grid visualisation, based on TileBars [14], which can be used for visualising interesting features in any type of linear data. The rows of the grids represent segments while each column represents a feature such as text transcripts or music. The colour (or grey-sale intensity) of each box encodes the presence or absence (or a score) of a feature at a segment. The hierarchical organisation using several nested levels of granularity allows the user to keep the sense of orientation while navigating towards ever finer levels of representation of the video.

3.2 Meeting Analysis

Meeting recordings are an example of multimedia documents in which complex communication patterns arise. For browsing of meeting recordings a specialised type of tools is typically employed: In addition to a video timeline these tools address elements of interest encountered in meetings which have previously been extracted from audio and/or video streams. These are for example: speaker activity, temporal display of speech transcripts, subjects of discussion and others. Each of the elements of interest is visualised in a separate interactive view enabling users to effectively browse meetings and play back parts of interest.

The Ferret Meeting Browser, developed within the M4 project [30], is a visual tool for browsing of speech segmentation and speech transcripts, which focuses on browsing a single meeting video [53]. After choosing which meeting and which data streams from a meeting shall be visualised, the user is presented with a graphical user interface consisting of a media player, and two scrollable timeline views for presentation of interval data streams.

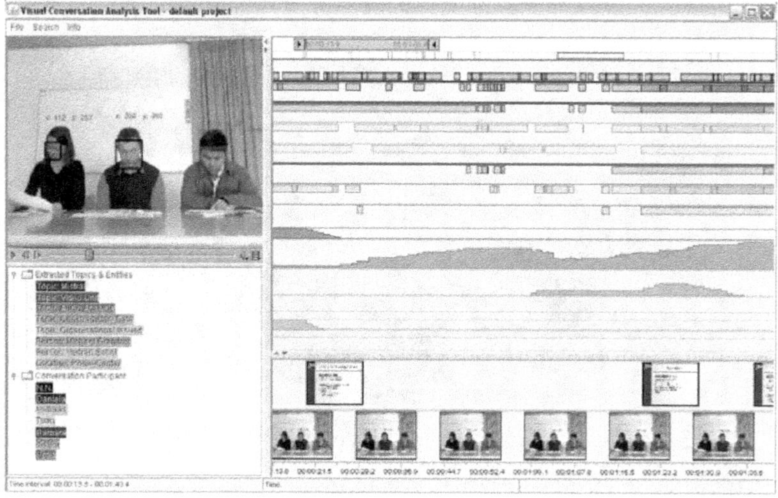

Fig. 4. Visual Conversation Analysis Tool (MISTRAL Project)

In [44], the Visual Conversation Analysis (VCA) tool (Fig. 4), developed within the MISTRAL project [32], is introduced. Conceptionally similar to the Ferret browser, the VCA browser offers several timeline views: An interval visualisation of speaker activity (up-right in the figure), a "topical intensity view" which visualises intensity of discussed subjects over time (below the activity view), a PowerPoint slide view (bellow the topical view), a timeline view of the video (bottom), and a temporal search result view (on top, above the activity view). Seamless temporal zooming and scrolling are realized through a time interval selection bar (the scroll bar-like horizontal component on the top of the GUI).

3.3 Browsing of Digital Image Repositories

Basic means for browsing collections of images are built into today's operating system's file managers: Microsoft Windows Explorer, iPhoto on MacOS or Konqueror [23] and Nautilus [34] on Linux provide a simple thumbnail view of images for a chosen directory. Dedicated tools such as Google's Picasa [35] offer more sophisticated methods for organising images but still adhere to the standard thumbnail paradigm. Formal experiments have shown that the application of certain visualisation techniques can improve the process of browsing and locating images of interest.

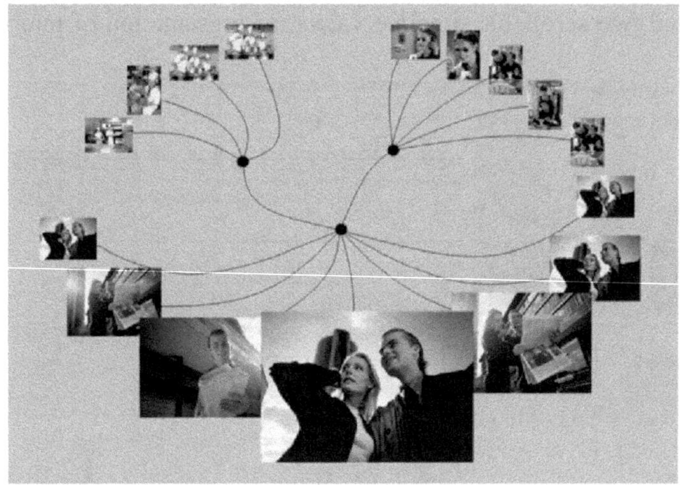

Fig. 5. Schematic of hyperbolic tree visualisation of digital images

FotoFile [24] is a system for managing digital media, such as photos, videos and audio files. Although FotoFile is capable of handling media with a temporal dimension (such as video and audio) the information is presented in a "photo-centric" way: a media object is always represented by image(s) (in case of a video these are extracted key-frames) with the corresponding sound and video material being attached to the image. User can annotate and search for materials, whereby media object properties are represented by a pre-defined set of metadata attributes, such as creation time, location, subject, people, title, description, etc. For browsing of multimedia data collections FotoFile offers (among other views) a Hyperbolic Tree visualisation [26] implemented using the software package from Inxight [18]. Hyperbolic Trees make use of hyperbolic geometry to visualise relations in large data sets which, if presented in a conventional way, would not fit onto the screen area. The data set is presented in such a way that information which is not of interest is pushed towards the edge of the view and shrank to make more place for information of interest (which is magnified). As the user changes the point of interest (i.e. the clicks on visualised objects) the geometry of the tree adapts automatically. Fig. 5 illustrates the principle of FotoFile's hyperbolic tree display built from different metadata types and their values.

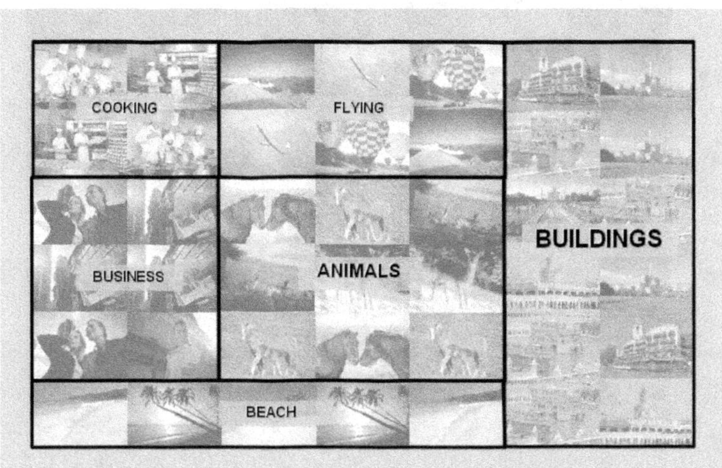

Fig. 6. Schematic of quantum tree map image layout

BubbleMaps and Quantum TreeMaps, which are variations of the TreeMap representation, are introduced in [5] and demonstrated within a PhotoMesa image browsing application (compare Fig. 6). In a Treemap, visualisation area is hierarchically partitioned into regions which have an area proportional to the amount of information entities they contain. In Quantum TreeMaps, as opposed to standard TreeMaps, the resulting areas do not have an arbitrary aspect ratio, but have dimensions which are integer multiples of the dimensions of an indivisible "quantum", in this case the size of a photo thumbnail. BubbleMaps are a further development of this concept in which the utilisation of the screen real estate is maximized leaving no or minimal empty space. BubbleMaps, as opposed to standard TreeMaps and quantum TreeMaps, can produce irregular, non-square areas.

3.4 Similarity and Relatedness

A similarity layout is a visual representation in which related objects are placed close to each other while unrelated objects are positioned further apart (i.e. spatial proximity is a measure for relatedness). In Fig. 7 a similarity layout produced by the Magick system [28] is shown. In this example images are positioned according to the similarity of their colour layouts. Other similarity measures, for example based on colour histograms, captions, metadata or even a combination thereof are also possible.

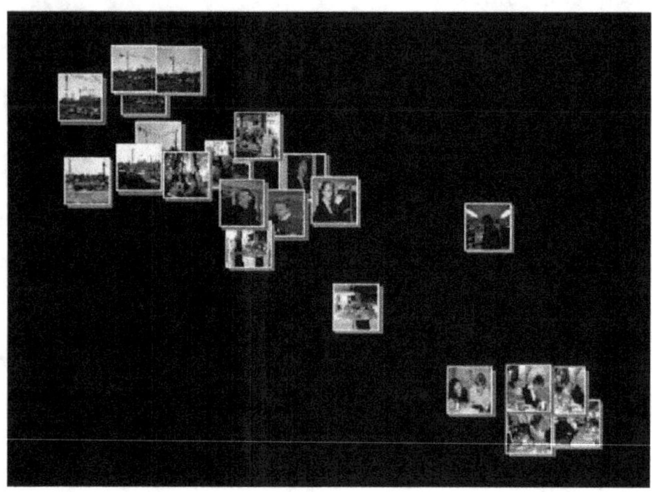

Fig. 7. A similarity layout of images

In [41] a technique for browsing a set of images based on a similarity layout of images is introduced. Similarity between two images can be computed depending on the similarity of image captions, or using a visual similarity measure which considers both the global image properties (i.e. colours and textures) and the spatial layout of image regions [42]. The computation of the layout was performed an incremental multidimensional scaling method described in [4]. Produced similarity layouts were evaluated with test users and their performance was compared to that of a random layout. It was found out that layouts based on visual similarity help the user to coarsely subdivide the data set into simple genres. However, a negative effect was also identified: some eye-catching images might be overseen in the similarity layout due to a "merging effect" of visually similar images. Layouts based on caption similarity were found helpful when browsing according to the meaning, provided available captions were of high quality.

Informedia Digital Library [49] is an advanced project at the Carnegie Mellon University aiming to develop techniques enabling full content search and discovery in large multimedia (primarily video) repositories. This was achieved through development and integration of different technologies such as automatic speech recognition, natural language understanding, extraction of key frames from videos, image processing and object recognition, and others. In the later stages of project development (Informedia II), due to the rising amounts of data, content summarisation and information visualisation techniques were developed [50], [51], [11].

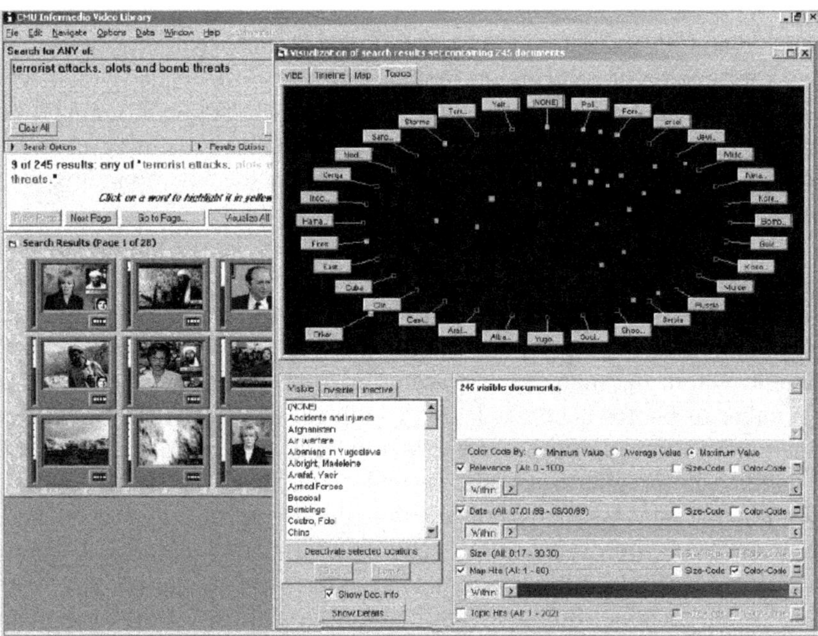

Fig. 8. Visualisation of video search results by topic. Taken from "Extracting and Visualizing Knowledge from Film and Video Archives" Presentation at the I-know'02, by Wactlar, H.D. (see [52])

In Fig. 8 (on the left) query results are visible as multimedia abstractions (i.e. summarizations through thumbnails). The height of the bar to the left of a thumbnail represents the relevance to the query, where the relevance to each query term is separately colour coded. In the same figure (top-right) a view for topical analysis of the result set can be seen: A number of extracted topics (points of interest) are initially placed on an ellipse (or on a circle) while query results, visualised as dots, are positioned between them. The proximity between a result and a topic is an (approximate) measure for the relatedness between the result and a topic. The user can rearrange the positions of points of interest by dragging them and watch the data elements adapt their position accordingly. Bellow the visualisation a number of filtering controls can be seen which allow the user to display only objects which satisfy certain conditions, for example objects from within a certain period of time or objects with at least a specified minimum relevance to the query.

3.5 Geospatial Analysis

Data which includes a geographical or, more generally, a spatial component is a natural candidate for being presented in a geospatial visualisation. Many multimedia documents do include geospatial metadata, which typically correspond to their place of creation. The already introduced Informedia browser offers a view where search results containing metadata providing a geo-spatial reference are visualised on the corresponding locations in a geographic-map.

In Tag Maps [20], a geo-spatial representation of a large number of photographs, is introduced. Similar to the previous example, results of a query are positioned on a geographical map depending on the geo-references which are accompanying each photograph. Each photo is represented by a rectangle within the map. Such a representation will become cluttered as the number of photos becomes large. To alleviate this problem two different summarisation approaches are available: The first one is based on spatial patterns in the photos and the second one is based on textual patterns describing the images (such as the caption, photographer information etc.)

The summarisation is based on the assumption that a majority of photos taken at a specific location concern something which is particularly interesting at that location. Summarisation algorithms utilise various metadata-based heuristics which are applied on spatial (where the photo was taken), temporal (when it was taken), textual/topical (tags, caption, and descriptions) and even social patterns (photographic behaviour of the user, or social network distance between the photographer and the user who is making the query). An evaluation revealed that summarisation algorithms performed well and the resulting visualisation was rated highly by test users. As a consequence geospatial analysis tools have already been integrated in online photo communities such as Flickr [13], BubbleShare [6] or Zooomr [56].

3.6 Motion and Trajectory Analysis

Spatial visual analysis can be used to asses the performance of player in sports. LucentVision [37] applies real-time analysis of video material collected by eight cameras to extract coverage and motion trajectories of the player and of the ball during a tennis game. The resulting visualisation provides precise information about performance and strategy of each player. Available visualisations include coverage maps (compare Fig. 9), motion trajectories, ball trajectories or service landing positions. The system has already been applied on hundreds of tennis matches.

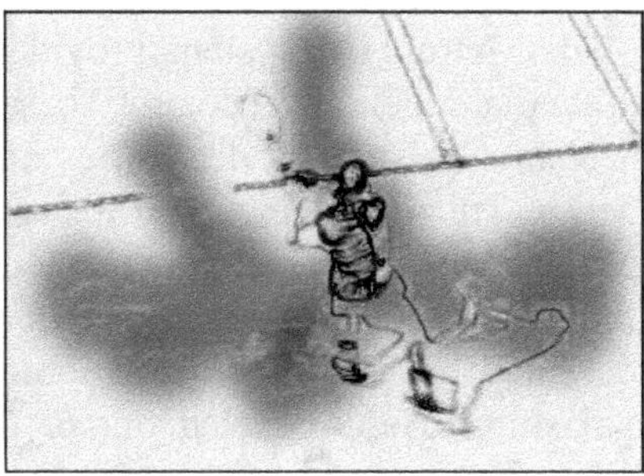

Fig. 9. Schematic of spatial coverage map for a tennis match

3.7 Navigating Social Networks

Another example of complex communication and social patterns are those arising from social networks in Web 2.0 based applications. Many of them, such as YouTube [55] or Flickr [13], deal with publishing, sharing and commenting of multimedia content. For exploration of these networks graph visualisation techniques are typically applied. It should be noted that graph layouting and visualisation is a very large, separate area of research, an overview as well as links to further reading can be found in [15].

In [1] an interactive graph visualisation is introduced to visualise relationships between music artists. Data on artists as well as relationships between similar artists are provided by AllMusic [2], an online music indexing and review site. In the graph representation, powered by TouchGraph [48], nodes represent artists where similar artists are linked by edges. Genre is encoded by colour so that clusters built around each genre can be clearly recognised. For navigation and exploration operations such as zooming, filtering as well as moving and expanding (to any desired depth) of nodes are supported.

3.8 Navigating Multimedia Repositories in 3D Virtual Worlds

Three-dimensional representation and virtual reality have nowadays become ubiquitous due to cheap and powerful 3D acceleration hardware. However, 3D information visualisation may suffer from problems such as occlusion and increased cognitive load on the user (as, for example, identified in [12]). To alleviate these disadvantages, the utilisation of and navigation in the third dimension is often "restricted" resulting in the so called 2.5D representations. Nevertheless, 3D user interfaces are common, especially because of their visual appeal and attractiveness.

MediaMetro [10] system provides an interactive 3D visualisation of multimedia document repository based on a city metaphor where information elements, such as directories and multimedia documents, are represented by suitable elements of a city such as buildings, blocks and streets. In MediaMetro directories are represented by blocks which are layouted by an algorithm based on quantum TreeMaps [5]. Multimedia documents are represented by buildings (i.e. cuboids). The facades of the buildings are visual summaries extracted out of the different constituent media types (such as keyframe images from videos or text summaries). As videos are the primary media type a representative frame will typically be placed on the roof, while the storyboard is placed around the building in the form of windows.

4 Conclusion

Although the examples given in this chapter provide an overview of only a small part of available multimedia visualisation techniques and systems, visualisation in the world of multimedia is not (yet) widely spread, at least not compared to some other applications areas such as information visualisation of text corpora or scientific visualisation. The causes for this are probably manifold and are not easy to identify. Visualisation of multimedia data introduces additional difficulties which are not present in, for example, textual data sets. Multimedia data are multi-faceted, i.e. include many different dimensions (or aspects) of information: topical, temporal and geospatial information, semantic metadata and structures, etc. Each information type might originate form a different modality where each modality has its own specifics. As a result applications using multimedia visualisation typically focus on a single dimension of the data, and even in cases where more than one dimension is addressed, each dimension is handled by a separate visualisation. This approach only allows for separate visual exploration and analysis of each dimension of the data, and fails to address cross-modal and cross-dimension patterns and relationships. Nevertheless, as we have seen in the majority of presented examples, this approach proved sufficient for performing tasks in focused, isolated application domains.

A visual tool capable of capturing the characteristics of multimedia data sets in their full extent has not yet (as of 2007) been proposed. Even if the area of information visualisation still does not offer means for achieving that goal, it offers a variety of visual representations which are still waiting to be applied on multimedia data. This is especially true for representations which attempt to address more than a single aspect of the data, such as those introduced in [21] (addressing geospatial and temporal dimensions) and [43] (addressing temporal and topical dimensions). Therefore, we conclude that multimedia visualisation is an area which is still emerging and which, despite its complexities and challenges, is a field offering appealing opportunities for research and innovation.

Acknowledgement

The Know-Center is funded by the Austrian Competence Center program Kplus under the auspices of the Austrian Ministry of Transport, Innovation and Technology (www.ffg.at), and by the State of Styria.

References

[1] Adamczyk, P.D., "Seeing Sounds: Exploring Musical Social Networks", Poster, In Proceedings of the 12th annual ACM international conference on Multimedia, 2004, pp. 512 – 515.
[2] AllMusicGuide. http://www.allmusic.com/, last accessed 06/2007
[3] Andrews, K., Wolte, J., Pichler, M., "Information Pyramids", IEEE Visualization, USA, 1997.
[4] Basalaj, W., Incremental multidimensional scaling method for database visualization. In Visual Data Exploration and Analysis VI (Proc. SPIE, volume 3643), January 1999.
[5] Bederson, B.B., "Quantum Treemaps and Bubblemaps for a Zoomable Image Browser", In Proc. User Interface Systems and Technology, pp. 71--80, 2001.
[6] BubbleShare, 2007, http://www.bubbleshare.com/, last accessed 06/2007
[7] Burkhard, R., Eppler, M. "Knowledge Visualisation", in: T.Keller and S.O.Tergan, "Knowledge and Information Visualisation", Springer, Germany, 2005
[8] Card, S.K., Mackinlay, J.D., Shneiderman. B., "Readings in information visualization: using vision to think", San Diego: Academic Press, pp. 1-34.
[9] Casares, J.P., "SILVER: An Intelligent Video Editor", Poster at CHI '01, In extended abstracts on Human factors in computing systems, pp. 425 – 426, 2001.

[10] Chiu, P., Girgensohn, A., Lertsithichai, S., Polak, W., Shipman, F., "Media-Metro: Browsing Multimedia Document Collections with a 3D City Metaphor", Proceedings of the 13th annual ACM international conference on Multimedia, 2005, pp. 213 – 214.

[11] Christel, M., Martin, D., "Information Visualization within a Digital Video Library", J. Intelligent Info. Systems 11(3), pp. 235-257, 1998.

[12] Cugini, J., Laskowski, S., "Design of 3-D Visualization of Search Results - Evolution and Evaluation", Proceedings of IST/SPIE's 12th Annual International Symposium: Electronic Imaging 2000: Visual Data Exploration and Analysis (SPIE 2000), San Jose, CA, 23-28 January 2000.

[13] Flickr, 2007, www.flickr.com, last accessed 06/2007

[14] Hearst, M.A., TileBars: Visualization of terms distribution information in full text information access. In Proc. of the ACM SIGCHI Conference on Human Factors in Computing Systems, p. 59-66, Denver, CO, May 1995.

[15] Herman, I., Melançon, G., Marshall, M.S., "Graph Visualization and Navigation in Information Visualization - A Survey", IEEE Transactions on Visualization and Computer Graphics, Vol. 6, No. 1, 2000.

[16] Hewett, T., Baecker, R., Card, S., Carey, T., Gasen, J., Mantei, M., Perlman, G., Strong, G., Verplank, W., "ACM SIGCHI Curricula for Human-Computer Interaction", Technical Report of the ACM SIGCHI Curriculum Development Group, 1992. http://sigchi.org/cdg/

[17] IDC, "The Expanding Digital Universe - A Forecast of Worldwide Information Growth Through 2010", Study available at http://www.emc.com/about/destination/digital_universe/pdf/Expanding_Digital_Universe_IDC_WhitePaper_022507.pdf

[18] Inxight Software, Inc., http://www.inxight.com, last accessed 06/2007

[19] SO 9241-11: Guidance on Usability (1998),
http://www.usabilitynet.org/tools/r_international_htm#9241-11, last accessed 06/2007

[20] Jaffe, A., Naaman, M., Tassa, T., Davis, M., "Generating Summaries and Visualization for Large Collections of GeoReferenced Photographs", Proceedings of the 8th ACM international workshop on Multimedia information retrieval, pp. 89 - 98, 2006.

[21] Kapler, T. and Wright, W., "GeoTime Information Visualization", Information Visualization, 4(2): pp. 136-146, 2005.

[22] Kienreich, W., Granitzer, M., "Visualising Knowledge Webs for Encyclopedia Articles", 9th International Conference on Information Visualization, UK, 2005.

[23] Konqueror, 2007, http://www.konqueror.org/, last accessed 06/2007

[24] Kuchinsky, A., Pering, C., Creech, M.L., Freeze, D., Serra, B., Gwizdka, J., "FotoFile: A Consumer Multimedia Organization and Retrieval System", Proc. ACM CHI99 Conference on Human Factors in Computing Systems, pp. 496-503, May 1999.

[25] Lalanne, D., Lisowska, A., Bruno, E., Flynn, M., Georgescul, M., Guillemot, M., Janvier, B., Marchand-Maillet, S., Melichar, M., Moenne-Loccoz, N., Popescu-Belis, A., Rajman, M., Rigamonti, M., von Rotz1, D., Wellner, P.,

"The IM2 Multimodal Meeting Browser Family", Joint IM2 Technical Report, March 2005.

[26] Lamping, J., Rao, R., Pirolli, P., "A focus+context technique based on hyperbolic geometry for visualizing large hierarchies", In Proceedings of the ACM SIGCHI Conference on Human Factors in Computing Systems (May I995), ACM.

[27] Long, A.C., Myers, B.A., Casares, J., Stevens, S.M., Corbett, A., "Video Editing Using Lenses and Semantic Zooming", Technical Report, http://www.cs.cmu.edu/~silver/silver2.pdf

[28] Lux, M., "Magick - Ein Werkzeug für Cross-Media Clustering und Visualisierung", Master's Thesis, Graz University of Technology, 2004.

[29] Lyman, P., Varian, H.R., How much information 2003, http://www2.sims.berkeley.edu/research/projects/how-much-info-2003/

[30] M4 Project, http://www.mistral-project.at, last accessed 06/2007

[31] Mills, M., Cohen, J., Wong, Y., "A Magnifier Tool for Video Data" CHI '92 Conference Proceedings, ACM Press, pp. 93-98, 1992.

[32] Mistral Project, www.mistral-project.at, 2006

[33] Müller, W., Schumann, H., "Visualisation Methods for Time-dependent Data – an Overview", Proceedings of the 2003 Winter Simulation Conference, Vol.1, pp. 737- 745

[34] Nautilus, http://www.gnome.org/projects/nautilus/, last accessed 06/2007

[35] Picasa, http://picasa.google.com/, last accessed 06/2007

[36] Pingali, G., Opalach, A., Carlbom, I., "Multimedia Retrieval Through Spatiotemporal Activity Maps", Proceedings of the ninth ACM international conference on Multimedia, pp. 129 - 136 , 2001.

[37] Pingali, G., Opalach, A., Jean, Y., Carlbom, I., "Visualization of Sports using Motion Trajectories: Providing Insights into Performance, Style, and Strategy", Proceedings of the conference on Visualization '01, Pp: 75 – 82, 2001.

[38] Ponceleon, D., Dieberger, A., "Hierarchical Brushing in a Collection of Video Data", Proc. 34th Hawaii International Conference on System Sciences 2001, pp. 1654-1661.

[39] Rennison, E., "Galaxy of News: An Approach to Visualizing and Understanding Expansive News Landscapes", ACM Symposium on User Interface Software and Technology, USA, 1994.

[40] Roberts, J.C., Wright, M.A.E., "Towards Ubiquitous Brushing for Information Visualization", Proceedings of the conference on Information Visualization 2006, pp: 151 – 156.

[41] Rodden, K., Basalaj, W., Sinclair, D., Wood, K., "Does Organisation by Similarity Assist Image Browsing?", International ACM SIGCHI conference, pp. 190-197, 2001.

[42] Rodden, K., Basalaj, W., Sinclair, D., Wood, K., "Evaluating a Visualisation of Image Similarity as a Tool for Image Browsing", Proceedings of the 1999 IEEE Symposium on Information Visualization, page 36.

[43] Sabol, V., Granitzer, M. and Kienreich, W., "Fused Exploration of Temporal Developments and Topical Relationships in Heterogeneous Data Sets", 3rd International Symposium of Knowledge and Argument Visualization, 11th

International Conference Information Visualisation. London, UK: IEEE Computer Society, 2007.

[44] Sabol, V., Gütl, C., Neidhart, T., Juffinger, A., Klieber, W., Granitzer, M., "Visualization Metaphors for Multi-modal Meeting Data", Workshop Multimedia Semantics - The Role of Metadata (WMSRM 07), Proceedings Band "Aachener Informatik Berichte", Aachen 2007.

[45] Shneiderman, B, Plaisant, C., "Designing the User Interface: Strategies for Effective Human-Computer Interactions", Addison-Wesley, Reading, USA, 2004.

[46] Tanaka, Y., Okada, Y., Niijima, K., "Interactive Interfaces of Treecube for Browsing 3D Multimedia Data", Proceedings of the working conference on Advanced visual interfaces, 2004, pp. 298 – 302.

[47] Thomas, J.J., Cook, K.A., "Illuminating the Path: The Research and Development Agenda for Visual Analytics", IEEE CS Press, USA, 2005.

[48] TouchGraph, 2007, http://www.touchgraph.com/, last accessed 06/2007

[49] Wactlar, H. D., Kanade, T., Smith, M. A., Stevens, S. M., "Intelligent Access to Digital Video: Informedia Project", Computer, 29(5), 46-52, 1996.

[50] Wactlar, H.D., "Multi-Document Summarization and Visualization in the Informedia Digital Video Library", in New Information Technology, 2001.

[51] Wactlar, H.D., "Extracting and Visualizing Knowledge from Film and Video Archives", Proceedings of I-Know'02 International Conference on Knowledge Management, 2002.

[52] Wactlar, H.D., "Extracting and Visualizing Knowledge from Film and Video Archives", Presentation at the I-know'02, available at http://i-know knowcenter.tugraz.at/previous/i-know02/downloads/hwactlar.pdf

[53] Wellner, P., Flynn, M., Guillemot, M., "Browsing Recorded Meetings with Ferret", Machine Learning for Multimodal Interaction, MLMI 2004, Martigny, Switzerland, June 21-23, 2004.

[54] Wellner, P., Flynn, M., Tucker, S., Whittaker, S., "A Meeting Browser Evaluation Test", Conference on Human factors in Computing System (CHI), Portland, Oregon, 2005.

[55] YouTube, 2007, www.youtube.com, last accessed 06/2007

[56] Zooomr, 2007, http://www.zooomr.com/, last accessed 06/2007

Automatic Extraction, Indexing, Retrieval and Visualization of Multimodal Meeting Recordings for Knowledge Management Activities

Christian Gütl

Graz University of Technology, Inffeldgasse 16c, 8010 Graz
cguetl@iicm.edu

Our knowledge society is faced with an increasing amount of data produced by a great variety of sources which results in diverse multimodal and multimedia data. Such data are characterized by a rich and complex structure in terms of intra- and inter-relationships. Their broad usage and reuse is limited unless sufficient technology-based methods for semantic annotation, extraction, indexing, retrieval and visualization can be developed. In this book chapter we address this problem domain by starting our investigation on general aspects of multimodal information systems and introducing a conceptual architecture for a generalized view about such systems. Based on this knowledge, we further focus on meeting scenarios as an example for a concrete application domain. This application area became increasingly important and a very active multidisciplinary research field over the last years. An overview about relevant research activities in the context of multimodal information systems is given. In the third part of this chapter we are discussing a concrete solution approach and practical experiences made in the MISTRAL research project.

1   Introduction

Our modern society is faced to cope with a huge amount of data, which has dramatically increased over the last decades. This may be illustrated by Lyman and Hal 2003, who emphasized in their study that *"Print, film, magnetic, and optical storage media produced about 5 exabytes of new*

C. Gütl: *Automatic Extraction, Indexing, Retrieval and Visualization of Multimodal Meeting Recordings for Knowledge Management Activities*, Studies in Computational Intelligence (SCI) **101**, 239–261 (2008)
www.springerlink.com

information in 2002." [23] The data output of a great variety of sources results in diverse multimodal and multimedia data, which are characterized by a rich and complex structure in terms of intra- and inter-relationships [39]. Human beings can not keep pace with this escalating amount of information and its relationships. Thus, its broad usage and reuse – including knowledge transfer and learning activities - is limited unless sufficient technology-based methods for semantic annotation, extraction, indexing, retrieval and visualization will be developed. Therefore, multimodal information systems gain increasing importance and have been an active research topic for years in many application domains, such as finance and economics [4], medicine and biology [5, 44], biometrics [14], human-computer interaction [31], communication science [38] and media research [44].

There is no doubt, supported media and modalities as well as systems' features and functionalities are strongly influenced by the application domains and the concrete application scenarios. Despite this diversity, however, a common set of conceptual units can be identified which forms above stated systems. In order to provide a generalized view of multimodal information systems (MIS) as well as to gain a framework for reviewing and describing systems in the remainder of this paper, an attempt is made to define a generic conceptual architecture in the following section.

2   CAMIS: A Conceptual Architecture of Multimodal Information Systems

Toward the course to a conceptual architecture for multimodal information systems, an investigation of the term 'multimodal' has resulted in a variety of opinions about its meaning; this may also be backed by [7]. Nigay and Coutaz split the term into its components and stated that 'multi' means 'more than one', and 'modal' may address the notions 'modality' and 'mode' [31]. Furthermore, they pointed out: *"The modality defines the type of data exchanged whereas the mode determines the context in which the data is interpreted"*. This allows a first abstract definition of a MIS as an information system which can deal with multiple types of data streams as input channel, interpret them in order to extract meaning or/and to convey meaning through multiple output channels. Already this simple definition implicitly includes aspects such as information extraction, information fusion and transformation as well as crossmodal information access. Based on this definition and by further building on insights gained in reviewing existing MIS in the above mentioned application domains, a generalized model of MIS built on conceptual units and important challenges is given:

- *Capturing:* This conceptual unit handles the provision of proper data streams from diverse sources for the further MIS process chain. Multi-modal data may include time series and sequence data (such as share prices, electrocardiograms or DNA sequences), temporally and spatially dependent media streams (such as audio and video recordings), and temporally independent data (such as textual documents). The most important challenge in this context is to provide data streams for further processing in a suitable form in terms of temporal and spatial granularity as well as in terms of content granularity. [13, 31]

- *Abstraction:* The conceptual unit for abstraction deals with data processing and information extraction. The level of abstraction may range from simple to more complex tasks. In increasing order it includes: (1) internally recorded data streams (such as video and audio streams for replaying), (2) compression or summarization (such as reducing video frame rates or capturing key frames), and (3) information extraction at multiple levels of abstraction (such as extracting color and shape features from video on a lower level, finding scenes and applying genre detection at a higher level of abstraction). Challenges in this context include the process of finding proper features and methods for inferring semantic meaning and mapping them to semantic concepts. [13, 31, 44]

- *Fusion:* In order to make use of the combination of multiple data streams, the fusion unit merges and combines information from the unimodal data sources. Again, this may be performed on different levels of abstraction, from fusion on a lower level of features to merging on a higher semantic abstraction level. Results of the fusion can also be utilized to improve unimodal extraction quality at the abstraction unit. Among other challenges, dealing with noisy and partly contradicting information in a fuzzy temporal manner is one of the most challenging tasks. Moreover, online processing of a huge amount of data for immediate feed-back is another challenging task. [5, 14, 31]

- *Storage:* This conceptual unit handles the persistent internal representation of the unimodal data streams and the extracted information. Furthermore, it must also manage the access and delivery of the data in a trustful and secure manner as well as take into account privacy aspects. Challenging aspects include the application of efficient data structures, metadata formats and standards as well as standardized information access. [8]

- *Retrieval:* In order to support the process of finding relevant information or delivering useful data, the retrieval unit manages browsing and searching in various modalities on different semantic levels and structure composites. Important challenges include crossmodal retrieval processes and combination of multi-modal information needs as well as the multimodal retrieval of temporally, spatially and logically structured information and the definition of suitable query languages. [6, 42]

- *Presentation:* This conceptual unit manages the combined and synchronized presentation of multimodal output data for information consumption. For the purpose of information browsing and search result presentation, it handles the visualization of annotated semantic information structures and the representation of Abstraction unit. Important challenges include crossmodal information presentation, combined and synchronized presentation of information corresponding to a specific modality and the handling of diverse logical abstractions and information granularities. [34, 38]

The conceptual architecture outlined above, what we term CAMIS (Conceptual Architecture of Multimodal Information Systems), gives an overview over the main units of a multimodal information system in order to support the understanding of such information systems on a domain-independent abstract level. It may also help to group user and system requirements logically on the basis of the identified conceptual units as well as to classify, assess and describe MIS on the basis of this proposed classification.

In the remainder of this chapter, we want to narrow down to our specific research topic which is multimodal information systems in the meeting application domain, what we termed 'Multimodal Meeting Information Systems' (MMIS).

3   Multimodal Meeting Information Systems (MMIS)

Face-to-face and virtual meetings have become increasingly important in our modern working live. According to Romano and Nunamaker managers and knowledge workers spend a surprisingly high amount (between 25% and 80%) of their working time in meetings [37]. Moreover, most organizations spend between 7% and 15% of their personal costs on meetings. Despite the high investment on resources, meeting results are unexpectedly often ineffective and inefficient. These results from various problems in the pre-meeting, in-meeting and post-meeting phases of the meeting life-cycle, see for example [20, 36]. To overcome the situation stated so far, research and development of technology-based support have been an active area since decades [32, 43].

An extensive literature survey has shown that over the last years research results as well as commercial applications and tools have emerged, which can sup-port a broad range of activities over the meeting life-cycle, see for example [21, 36]. The aim of the 'pre-meeting phase' is to prepare the meeting, which includes activities such as reviewing previous meetings, specifying the meeting goal, preparing necessary documents, selecting meeting members and managing schedules and resources. The 'in-meeting phase' aims to ensure an efficient and effective performance of the meeting. It includes moderation and guidance activities as well as monitoring and recording activities. The aim of the post-meeting phase is to analyze the meeting and infer next steps as well as to ensure, that relevant information is delivered and accessible to meeting participants and other stakeholders.

From the knowledge management viewpoint, interesting and important knowledge is discussed and even new knowledge is created in meetings, which is worth to capture, store and make accessible for further usage [12, 32]. Consequently, support for knowledge transfer and learning activities in this context is a further key issue.

Multimodal meeting recording, processing, retrieval and access, what we term a multimodal meeting information system (MMIS), has gained increasing interest in the last few years. In order to provide an overview how MMIS can support activities over the meeting life-cycle, an overview over interesting and important projects is given in the following sub-chapter.

3.1 MMIS Research and Systems at a Glance

Multimodal meeting information systems have been an active research topic for a long time. To give one example, the project NICK had already emerged in the mid 1980s. From the MMIS point of view, the concept consists of capture, analysis and presentation components. The technology-enhanced meeting environment is built on an electronic blackboard system, interconnected personal computers for each meeting participant and soft-ware for meeting facilitation, such as agenda information as well as support for communication and group memory access. Captured modalities include collaborative notes, participants' notes and votes, and audio and video recordings as well. Analysis is mainly focused on meeting statistics to quantify aspects of meeting effectiveness, and the presentation component provides access to the meeting information for further usage. [11]

Over the last years the multidisciplinary research on the meeting topic has gained increasing interest in diverse research domains, such as automated audio and video signal processing as well as in human-human and human-computer interaction. The importance of the meeting application domain may be reflected by several interesting and partly innovative research projects in Europe and USA.

The European Commission (EC) funded project M4 (multimodal meeting manager) started in 2002 and had a duration of three years. The aim of the project was to research and develop a demonstration system which enables structuring, browsing and querying of an archive of automatically analyzed meeting recordings. By focusing on MMIS aspects, the meeting room was equipped with sensors for multimodal data capturing, such as multi-channel audio and video data, textual information (agenda, discussion papers, text of slides), and interaction stream, which can support analyzing events within the meeting (for example, mouse tracking from a PC-based presentation or laser pointing information). The audio-based metadata extraction includes speech-to-text transformation, speaker detection, segmentation and tracking. The video-based metadata extraction includes face-detection, person tracking and person action recognition, such as gestures and actions. Fusions of extracted information from different modalities enable a multimodal identification of intent and emotion as well as multimodal person identification. These extracted information and automatic annotations are used to integrate and structure captured modalities in a meeting archive and makes it accessible for further usage. [24, 25]

Similar to the M4 project, the AMI (Augmented Multi-party Interaction) research project, a EC-funded integrated project, was started in 2004 and aimed in building systems to facilitate meetings and their documentation [1]. The EC-funded research project AMIDA (Augmented Multiparty Interaction with Distance Access) has been started in 2006 in order to continue the research of the AMI project. The AMIDA project aims in shifting meeting recordings to more challenging tasks in live meeting support with remote participants by using affordable commodity sensors, such as webcams and cheaper microphones. Scientific challenges include real-time processing and processing of lower quality audio and visual signals. [2]

Another interesting aspect in technology-enhanced meeting systems is the proactive assistance and support over the meetings life-cycle. CHIL (Computers in the Human Interaction Loop), a EC-funded integrated project started in 2004, aims to research and develop environments in which computers support humans who focus on interacting with other humans. The application scenario is focused on situations in which people interact face to face with people, for example exchange information, collaborate to solve problems jointly, learn, or socialize. The proactive support requires online-processing and real-time interaction which is more challenging in processing input modalities such as speech and language, gestures, body posture, and other data streams. [10] Another ambitioned research project is CALO (Cognitive Assistant that Learns and Organizes), which was initiated by the US Defense Advanced Re-search Projects Agency (DARPA). The main goal is to research and develop a cognitive agent-based system that can reason, learn, and respond to assist in military situations. Meeting support is covered by the Task Discussion Component, which focuses on a meeting understanding by observing users' multimodal dialogue during the meeting. Modalities of interest are speech, pen input, facial expression, gesture, and body movements. Indented assistance includes meeting summarization and active suggestion of relevant documents. [9]

3.2 Meeting-specific Findings from the CAMIS viewpoint

Research, design and implementation activities in the scope of MMIS as outlined in the previous section have shown ambitioned approaches, promising results but also big challenges. In this sub-chapter important findings from these research projects and from our literature survey are given from the viewpoint of the conceptual architecture of multimodal information systems (CAMIS).

Capturing

Over the life-cycle of meetings a great variety of valuable input channels is worth to be taken into consideration, and might be useful for information extraction and further usage as well. This section is based on [1, 2, 3, 9, 25, 36].

In the pre-meeting phase, meeting agenda and meeting participant lists are of particular interest. Additionally, meeting-relevant background documents such as project documentations, meeting minutes of previous meetings or other projects, and information about meeting participants are addressed by research work in that context.

In the meeting phase, multiple video and audio channels are the most prominent information sources for face-to-face and virtual meetings. Most of the research projects also take into account the information of presentation devices (interaction with presentation computer, electronic blackboard and the like) and agenda status. Research projects also capture information about public and private note taking, voting, environmental conditions of the meeting room, and physiological information about meeting participants.

In the post-meeting phase, mainly hand-made or revised computer-generated meeting minutes are exploited as a further input channel. Furthermore, some re-search projects enable the annotation of meeting re-cordings and extracted information (for example assigning description about topics and activities), which can also be very useful modalities for further usage.

Types of input channels and data sources are data streams of diverse media and time series, which are linked with temporal and spatial information as well as time-dependent and time-independent textual information. Important challenges on technological level include the provision of data of proper granularity (temporal and spatial resolution) and quality (accuracy and high information/ noise ratio) for further information processing and extraction at the abstraction and fusion units (see also below). There is a field of tension, however, between the needs of high quality sensors (such as high resolution video cameras and directional microphone) and available devices in practical meeting setups. Additionally, at the organizational level security and privacy problems need to be solved.

Abstraction and Fusion

MMIS deal with diverse abstractions of captured input channels and their fusion on various semantic levels, which also may be backed by [1, 3, 9, 25].

On the lowest level, captured data are recorded and converted into representation for further internal processing and access for external usage. Fusion tasks are mainly focused on synchronization and merging of multiple data streams without higher semantic meaning. Illustrative examples are audio and video format conversation into streaming media formats, synchronization of multiple audio and video data streams, and synchronized merging of multiple audio streams into one audio track.

Towards the course to semantic information extraction, the next abstraction level deals with feature extraction and fusion on low semantic meaning. To illustrate this, speech signal processing extracts phonemes and frequency spectrum, video signal processing operates with spatial histograms, and text processing extracts multidimensional description vectors and named entities. An example for fusion on low semantic meaning is the combination of multiple speech levels and sound signal delays given by a microphone array.

On a higher abstracted level, semantic information is derived from extracted features or features are mapped to semantic concepts. From sound data streams, different sound sources (such as mobile phone ringing, clapping hands and laughing) and speech-to-text transcripts are extracted. Additionally, speaker activities, spatial location and speaker identification are topics of interest within research projects. From video data, face recognition, facial expressions, body movements and body language, movements and meeting participant detection is addressed in the meeting research arena. Textual information, such as speech-to-text transcripts, notes during the meeting and agenda are processed in order to provide abstracted information. Examples are summaries as well as meeting minutes and topics addressed in the meeting. Fusion on a higher semantic level enables to identify contradictions or to boost the confidence of semantic information derived from lower level features. To illustrate this, speaker identification and localization can be processed on audio and video input data, and the final results including a confidence level is given by the fusion of information from both modalities. Moreover, the fusion of diverse semantic information extracted and the observation over the timeline enables to derive further semantic information on a higher abstract level. Research projects in the meeting domain focus on detecting meeting actions (such as monologue, discussion, presentation, note taking and voting), roles in meetings (such as moderator and presenter), argumentative and emotional states, communication structures and human-human interaction, and high level meeting segmentation (such as opening and closing of a meeting). Additionally, statistic information about meetings is also part of the scope of meeting research projects, addressing meeting duration, activity of meeting participants, topics addressed in meetings, and the like.

Extracted semantic information may be used to annotate meeting recordings automatically, to enrich information about the meeting and to enable meeting segmentation, retrieval and access from diverse points of view. Important problems in the context of abstraction and fusion include 'noisy' data (such as echo effect and ambient noise in sound data streams and changing light conditions in video data streams) and the fuzziness of temporal and spatial mapping of events from different modalities (such as speaker location by audio and video sources). Methods for basic information extraction on particular aspects are still very error prone, such as speech-to-text transcripts in typical meeting scenarios. Generally speaking, the automatic extraction for most of above mentioned information on a higher semantic level is still not solved satisfactorily and provides much room for further research and development activities.

Storage

This unit handles the persistent storage and the delivery of the representations of captured unimodal data streams and extracted semantic information. Relevant literature for this conceptual unit includes [8, 21, 35].

Meeting information archives are built on the file system or based on database management systems. They are designed to manage and deliver captured audio and video modalities as synchronized data streams. Additionally, such components deal with temporal information (time marks) for other modalities (such as interaction events with presenter PC, agenda, and speech-to-text transcript) for synchronization purposes and time-based access. Furthermore, meeting-relevant documents or at least metadata about these documents are managed by the storage component.

From the interoperability point of view, metadata formats used in meeting re-search projects do not sufficiently address standards, such as MPEG-7 and MPEG-21 as for example described in [26]. This seems to be a major disadvantage for further internal and external usage, for example the application of pre-existing indexing and retrieval systems or the integration into external information systems. In order to ensure security and privacy, a fine-grained user role model and access right policies are required, but these aspects are not adequately addressed in meeting research projects yet.

Retrieval and Presentation

Captured modalities of meetings and extracted semantic information need to be accessibly based on users' information demands. Therefore, the retrieval and presentation units in the meeting application domain manage the indexing process, enable browsing and searching capabilities, provide approaches and metaphors for information visualization and handle the multimodal interaction with users. This section is based on [21, 22, 35].

Extracted metadata and textual information from documents are the base for the indexing and retrieval process. Meeting research projects mainly focus in this con-text on speech-to-text transcripts, agenda, and speaker segments. Additionally, meeting-relevant background documents and meeting activities are of partial interest. In principle, any extracted information at various semantic levels can be utilized for the retrieval process, but this would also increase significantly the complexity of such systems. From a general viewpoint, indexing and retrieval of temporally, spatially and semantically interrelated and hierarchically structured in-formation is a very challenging task and still offers much room for further re-search.

The presentation unit, which is also termed meeting browser in meeting re-search literature, is responsible for the visualization of search results, extracted semantic information and captured data streams of meetings. Various developments within meeting research projects mainly focus on speech-to-text transcripts and speaker segmentation visualization combined with synchronized video and audio streams. Some of the developed systems also present the content of presentation slides or electronic blackboards, statistical information about speaker activities, speaker duration and the like. Ideas for accessing meetings from different viewpoints (such as employee roles, meeting participants and absentees) are outlined by some of the research projects, however project results and actual prototype implementations do not cover users' demands appropriately.

Last but not least an interesting insight from the knowledge management point of view is given. Despite the broad interest on technology-based meeting support and the significant number of diverse meeting research projects, concrete project outcomes result in applications, which mainly remain isolated from other business information flows. To overcome this, a close integration into other business processes is still an important objective and research focus. Interesting and important examples of such business processes are knowledge transfer, decision making and learning activities. This may also be backed by [18, 12]

4   The MISTRAL System

The aim of this section is to introduce our research and to discuss interesting findings on multimodal information systems within the MISTRAL project, a two-year research project nationally funded by the Austrian government, which started in 2005. The abbreviation MISTRAL stands for "Measurable Intelligent and Reliable Semantic Extraction and Retrieval of Multimedia Data". The motivation for the MISTRAL project was to research and develop novel methods to cope with the huge amount of multimodal data by exploiting semantic meaning and the complex structure in terms of intra- and inter-relationships. [18, 29, 39]

Fig. 1 outlines the general, application-domain independent MISTRAL architecture at a glance and shows the relations to the CAMIS units. On the outer left side of the picture, the unimodal units for audio, text, video and other environ-mental sensors are dealing with capturing and abstraction within the CAMIS model. Based on the unimodal data processing, the Multimodal Merging unit com-bines extracted data on various abstraction levels based on semantic, spatial and temporal characteristics. Further information enrichment and contradiction checks on extracted information are performed based on domain knowledge within the Semantic Enrichment unit. Both, the Multimodal Merging unit and the Semantic Enrichment unit address fusion tasks of the CAMIS model. The captured data from different modalities together with metadata as well as extracted and derived information on various semantic levels are managed and made accessible by the Data Repository unit. These units stated so far assemble the MISTRAL core system.

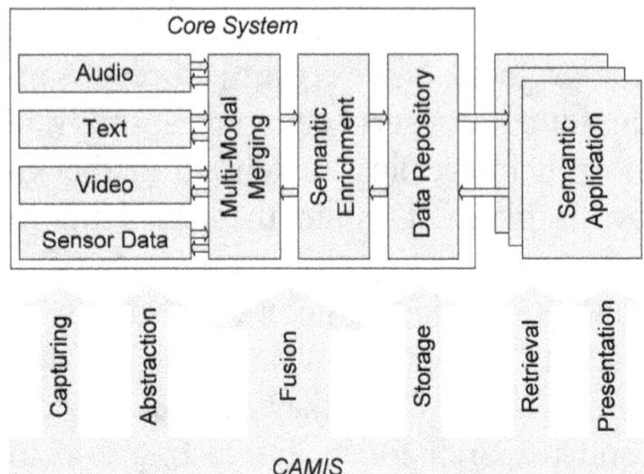

Fig.1. MISTRAL overall architecture and its mapping to the CAMIS units.

From the information flow point of view, it is worth to mention that different types of units have information feedback loops. This approach enables for example unimodal units to make use of extracted information of other units in order to improve semantic extraction performance or to train und improve extraction models. Other units in the core MISTRAL system can take advantage similarly. In order to gain high flexibility in terms of extensibility and interchangeability of methods and services, web services will build the basis for such an infrastructure as they are designed to support interoperable machine-to-machine interaction over net-works.

Different semantic applications can make use of the functionality of the core system and the great variety of available data on different semantic levels. Based on concrete application scenarios different aspects of retrieval and presentation related to the CAMIS model are addressed.

The consortium of the MISTRAL research project has decided to focus on the face-to-face meeting domain for the concrete sample application, which was motivated by an increasing interest in research community and commercial organizations. The MISTRAL system from the meeting viewpoint is discussed in the following sections.

4.1   The MISTRAL Core System

This section is based on the research work from [18, 39]. In order to increase the understanding of the meeting-specific MISTRAL core system, the meeting setup for face-to-face meeting capturing and relevant data sources are briefly outlined. Up to 4 meeting participants can sit in front of a table. In opposite of the meeting participants one high-resolution video camera and a microphone array (built of 4 microphones for far-field recording) are the sources for the video and audio capturing units. Additionally, data about the interaction with the presenter PC is the source for the sensor capturing units. Finally, meeting-relevant background data from document and information repositories (such as agenda, meeting participant information and project information) provide additional data streams for the textual capturing unit.

From the abstraction point of view, MISTRAL's unimodal units extract features and derive semantic information at various semantic levels (see below). The aim within this book chapter is to give an overview over the different types of extracted semantic information. Technical details and methods for basic feature extraction can be found elsewhere in (MISTRAL).

The video unit applies a generic, robust face detection and localization method for meeting participant identification as well as spatial localization and tracking. Thus, actions such as stand-up, moving and sit down can be detected. Video signal processing requires a high resolution video stream and constant light conditions for sufficient face recognition.

The audio unit processes multiple far-field microphone signals and focuses on voice activity detection, spatial identification and tracking of audio sources, and speaker indexing. The quality of processing result is influenced by ambient noise (such as fan noise from beamer, computer or air conditioner) and echo signals. Based on the audio signal and speaker segmentation, speech-to-text extraction is performed by applying a tool from (SAILLABS). This method enables a speaker-independent processing on domain knowledge specific phoneme and dictionary databases. Our experiences have shown that speech-to-text extraction is very error prone caused by ambient noise and echo signals from far-field microphone signals and meeting participants without proper speech training. To improve the performance, further information about the meeting (such as presentation slides, agenda, project information) need to be taken into account and based on that, a tailored phoneme repository and dictionary must be applied.

Within MISTRAL's meeting application, the sensor data unit exploits the data stream from meeting participants' interactions with the presenter PC. This modal-ity includes information of each keyboard and mouse event, active application and interaction with objects and documents. For example, this module provides information about the active slide of a PowerPoint presentation or a Web site presented at the meeting. This information can be further exploited by applying text processing tasks as described in the next paragraph.

The text unit focuses on natural language processing and text mining on textual meeting information. It includes extraction of named entities, supervised concepts and unsupervised concepts, such as participants' names, company names, project relevant keywords and associated terms. Besides common known problems in natural language processing and text mining, multilingual meeting documents (for example project information in English language and speech-to-text extraction in German language) have caused additional extraction problems. A possible approach for further performance improvements is to apply meeting or project specific domain knowledge and domain ontology.

The multimodal merging unit is responsible for the temporal and spatial fusion of extracted meeting information and events on various semantic levels. This unit can either merge information from different unimodal units with similar semantic meaning or derive new information by combining information and events from different units. It is in particular challenging to merge fuzzy temporal and spatial co-occurrence of information from diverse modalities. The semantic enrichment unit handles the dissolving of identified conflicts caused by contradictory information extracted from different units. Additionally, it focuses on the identification of actions on a higher semantic level, such as discussion, presentation, and the like. In particular model building for such actions is very complex and requires the ability to compromise settlements. For example, it is not clear how to model the difference between a short interruption in a presentation and a discourse. Both, Multimodal Merging unit and Semantic Enrichment unit, deliver information for further usage and access in a MISTRAL-specific XML data format or it can also be transformed to the MPEG-7 metadata standard.

The data repository for MISTRAL's meeting application is responsible for the persistent storage, management and access for further usage by taking into account access rights on user and group level. This includes the storage of the audio und video recordings and its delivery as real media streaming media files. It also comprises the management of meeting relevant documents and meeting participant in-formation. In addition to this, the data repository also handles the semantic meeting information extracted from the unimodal and mulitimodal units, most of them enriched with temporal information and references to the meeting recordings. Not only access rights are important for this unit, but also privacy aspects and legal problems may become an issue in practical applications.

4.2 The MISTRAL Semantic Application

MISTRAL's semantic applications in the meeting scenario are motivated by the lack of multimodal meeting information system integration for other business processes, knowledge management and learning activities in organizations. It comprises a meeting information retrieval and meeting information visualization unit. This section is based on [15, 16, 18, 19, 28, 39].

The meeting information retrieval unit is based on the open source search system xFIND, which is built on three services: (1) the Gatherer requests, filters, retrieves and pre-processes documents in order to provide a configurable set of metadata for further processing, (2) the Indexer caches the metadata, builds on top of that indexes and post-processes search results for presentation, and (3) the Broker provides the interface to the user and prepares the search results for information delivery. Further information can be found for example in [17].

The MISTRAL retrieval system uses a specialized gatherer for retrieving and processing data from the meeting repository. In the current version of the Retrieval unit, the following data sources are processed: (1) metadata about the meeting such as meeting title, place and time, meeting duration, participants and related project. (2) Extracted semantic information such as speech-to-text excerpts, active speakers, visible meeting participants and presenter PC interaction. (3) Meeting relevant documents such as agenda, presentation slides and other related back-ground documents, for example project description, design documents and the like.

Unlike other multimodal meeting information systems, the approach within the MISTRAL project focuses not only on the temporal and spatial indexing of diverse modalities but also on different logical meeting segmentations for the retrieval process and the search result presentation. The left side of Fig. 2 may illustrate this: the MISTRAL retrieval system supports four different granularities, (1) the entire meeting, (2) segmentation based on speaker activity, (3) topics according to the agenda of the meeting, and (4) segmentation based on the granularity of presentation slides. This enables not only to search over an entire meeting or over a specified time period, but also to search for particular segments and restrict the search results based on these segments.

To give an example, one can search for the term 'design' in the speech-to-text excerpt and restrict the search only to results also occurring within agenda topic "open issues". Different search forms provided by an adapted xFIND Broker service supports the users to describe their information needs. Power user gets the freedom to construct complex search queries by applying Boolean expressions, specifying specific indexes and granularities in an expert search form. A simple search form hides this complexity and enables users to search over any indexed information on the granularity of entire meetings by typing only one or several keywords.

Specific forms support the users to search for example on the granularity of speaker segments, as depicted on the right side of Fig. 2.

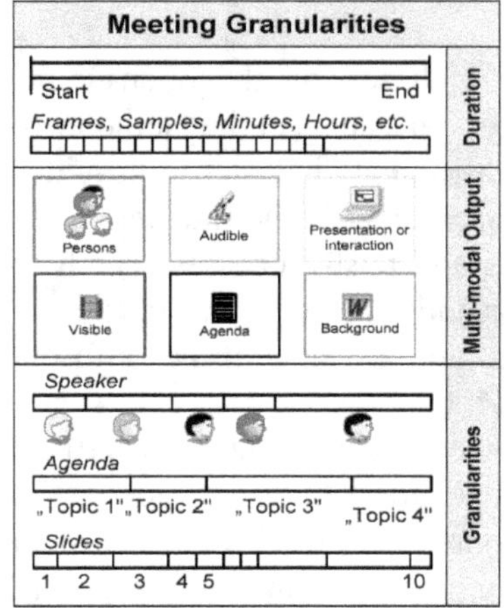

Fig. 2. Meeting information granularities for indexing and retrieval (left picture with permission from V.M. García-Barrios taken from [16]).

A further distinguishing feature of the MISTRAL retrieval system is the delivery of meeting information from different modalities for each search result corresponding to one of the four meeting granularities. The modalities of interest and their extent of information delivered can be specified within the search request. Availability and extend of information for each modality is determined by the above mentioned gathering and indexing process. To illustrate this comprehensive information delivery, one example of a search result presentation from the adapted xFIND Broker is depicted in Fig. 3. This specific search result view lists relevant speaker segments from meetings. It provides to the users an overview over the related meeting and for each of the meeting segments descriptive metadata, a relevant text sniped from the speech-to-text excerpts and meeting participants. It also renders thumbnails from the meeting scene temporally linked to the speaker segments. Additionally, information is delivered about the segment-specific meeting document actively used by the presenter PC. Further information can be requested by clicking on hyperlinks, such as the speech-to-text excerpt or the meeting recording about the specific segment.

The search-engine-like result presentation described above supports users to search for specific meeting information and explore and access search results in an easy way. However, in order to get a more comprehensive overview and to enable information discovery, a further result presentation in form of various visualization metaphors has been implemented. This meeting information visualization tool acts as a xFIND Broker, i.e. it provides the interface to the users and communicates with the MISTRAL Meeting Information Retrieval unit. The visualization tool is built on two server-side components, the application-dependent data Pre-processors and the data Normalizer, which are responsible for data handling and data provision for the visualization process. The Preprocessor compiles a search query in order to specify the user's need of information, granularity of meeting in-formation as well as modalities of interest and their extent. The retrieved search results are pre-processed according to a specific user problem; for example to build a vector space model or to compute statistical data. In the next step, the resulted data are converted by the Normalizer in order to fit the normalized data structure of the Visualization Client.

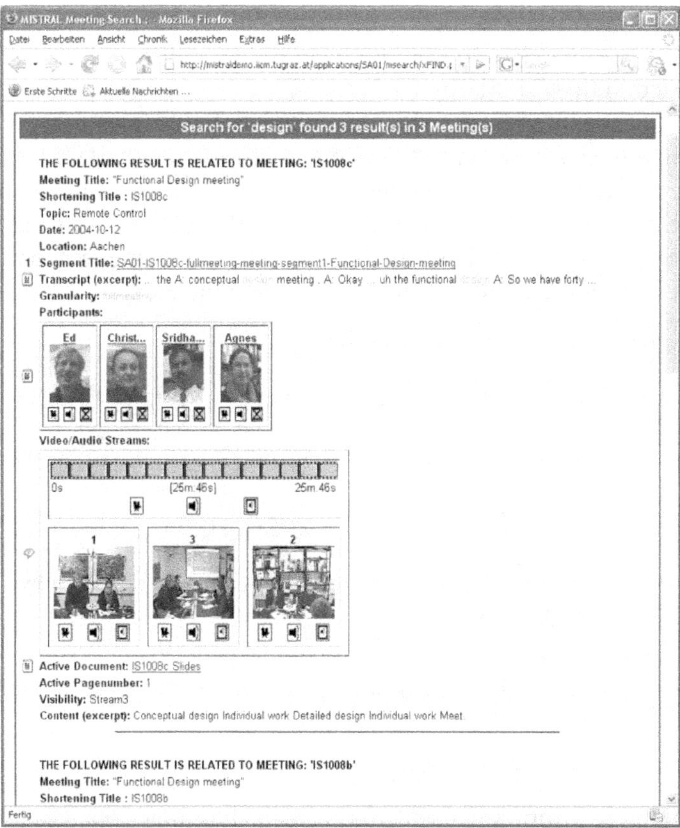

Fig. 3. Comprehensive meeting information presentation for search results on the granularity of speaker segments.

The Visualization Client has implemented some general management components, such as data handling and caching, communication infrastructure and an ob-server for synchronization of different visualization views. The client can either be used as stand-alone Java application or it can also be integrated in a web application as a Java applet. Because of a well-defined interface, various visualization tools can be easily implemented and integrated. Three generic panels - Filter and Control Panel, Linear List Panel and Info Panel – have been implemented which support data handling and inspection. For information visualization and information discovery, currently two diverse metaphors are useable: (1) The Scatter Plot view (see left side of Fig. 4) uses ordinate, abscissa, icons or pictures and their size to visualize 4 diverse dimensions of the multi-dimensional data in a two-dimensional plane. Users are free to choose any dimension from the data vector and assign them to one of the four visualization dimensions. (2) The VisIsland view (see right side of Fig. 4) is based on the visualization principle to place graphical representations of data vectors in accordance to their similarity (for ex-ample by using the Cosine similarity coefficient calculated from values of metadata fields related to the specific user problem) in a two-dimensional plane. Consequently, similar data vectors are spatially grouped together and form islands of similar objects in the graphical view. Both visualizations enable users to gain insights about aspects of one or several meetings, such as activity of meeting participants, frequency of topics addressed in meetings, and the like.

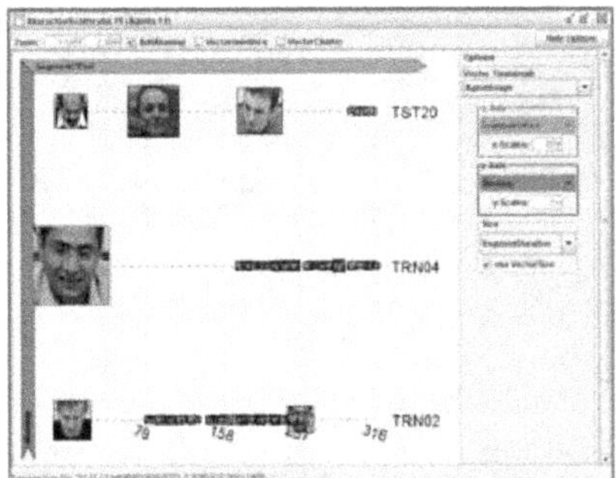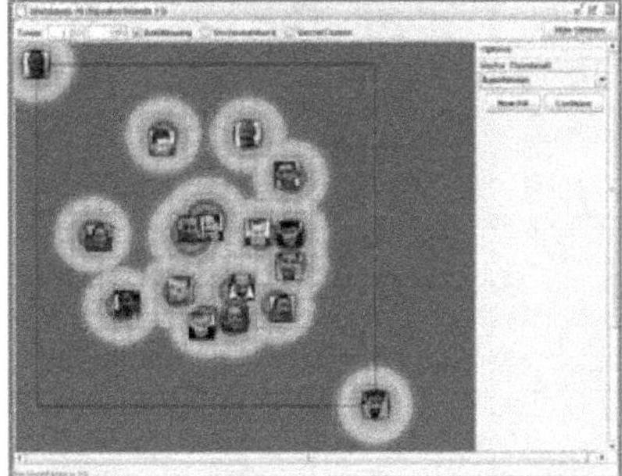

Fig. 4. Meeting information visualization based on the scatter plot metaphor (left side) and the VisIsland methaphor (right side).

As already outlined in the end of section 3.2, a close integration of multimodal meeting information systems (MMIS) into other business processes is still an important objective and research focus. In order to contribute to this research topic, we have designed and experimentally implemented a solution for a close integration of the MISTRAL system with other business applications, such as workflow systems, knowledge management systems and learning management systems. Fig. 5 illustrates the basic idea. Information systems deliver specific content in accordance to user's and groups' roles. Based on the delivered content and context (user role, group role, task and the like) customized related concepts are also selected by the Concept Modeling System and delivered via the information system to the user client. For each concept one or several information requests are rendered in the user client as hyperlinks in a navigation-like section in the client window. Thus users are enabled to retrieve information of interest from the MISTRAL system by just clicking a hyperlink. This approach supports users to use multimodal meeting information systems within their daily working environments.

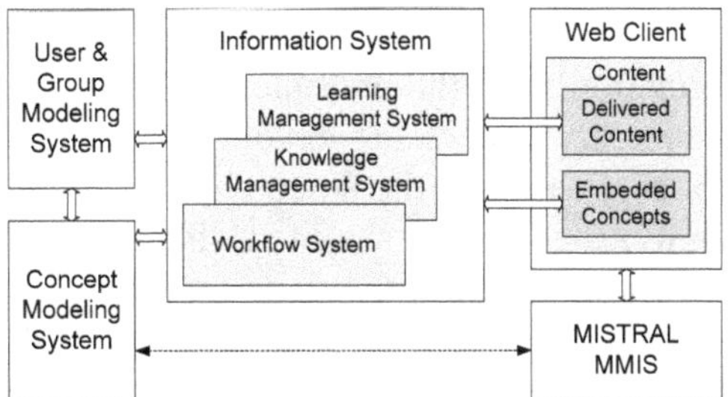

Fig. 5. Concept-based access to the MISTRAL multimodal meeting information system (MMIS) from users' daily working environments.

5 Conclusions and Outlook

In this chapter, we have given an overview over the automatic extraction, indexing, retrieval and visualization of multimodal data for knowledge management activities.

First, we have shown the importance of extracting and using semantic information from multimodal data in diverse application domains and application scenarios. As a result of the literature survey, a Conceptual Architecture of Multimodal Information Systems (CAMIS) has been developed. Six main conceptual units have been identified, namely (1) Capturing, (2) Abstraction, (3) Fusion, (4) Storage, (5) Retrieval, and (6) Presentation.

Secondly, as an interesting but also challenging application domain, the motivation for technology-enhanced meeting support, in particular meeting information recording, archiving, retrieval and accessing has been discussed. Existing multimodal meeting information systems have been reviewed, and based on the CAMIS model, main functions and features of modern meeting information systems have been outlined.

Finally, the MISTRAL research project in the context of the meeting application domain has been introduced. Furthermore, the MISTRAL architecture has been linked to the CAMIS model and functions of the 6 conceptual units have been outlined. Additionally, an approach has been introduced for linking meeting information into employees' daily working environment to support knowledge transfer and learning on the job activities.

For future work, we intend to evaluate the application of meeting information for corporate learning activities in a real-life setup. Furthermore, we want to research enhanced methods for speech-to-text transcripts by taking into account speaker characteristics, knowledge domain of the meeting and information from the presenter PC. We are also interested in applying the system for processing recorded lectures (e.g. MIT open courseware repository) to make them searchable and accessible on a semantic level. Finally, together with a big software developing company, we are currently working in the early planning stages of a concept how to integrate parts of our research results with the improved archiving and retrieval features of the company's virtual meeting tool.

Acknowledgement

The project results presented in this paper were partly developed within the MISTRAL research project. MISTRAL is financed by the Austrian Research Promotion Agency (http://www.ffg.at) within the strategic objective FIT-IT (project contract number 809264/9338). The support of the following IICM members is gratefully acknowledged: Victor Garcia, Helmut Mader and Martin Ruhmer.

We also gratefully acknowledge the following resources for meeting data provision: (1) The AMI Meeting Corpus, and (2) I. McCowan, S. Bengio, D. Gatica-Perez, G. Lathoud, F. Monay, D. Moore, P. Wellner, and H. Bourlard, "Modeling Human Interaction in Meetings", in Proc. IEEE Int. Conf. on Acoustics, Speech and Signal Processing (ICASSP), Hong Kong, April 2003. For further information see also http://www.m4project.org and http://mmm.idiap.ch.

References

1. AMI (2006) AMI overview and prospects for future research. State of the Art White Paper, January 2006, last retrieved 2007-03-26 from http://www.amiproject.org/ami-scientific-portal/documentation/annual-reports/pdf/AMI-overview-and-prospects-for-future-research-Jan2006.pdf

2. AMIDA Overview. IDIAP Research Institute, last retrieved 2007-06-23 from http://www.idiap.ch/speech-processing.php?project=102

3. AMIDA (2007) State-of-the-art overview: Localization and Tracking of Multiple Interlocutors with Multiple Sensors. Technical Paper, Jan. 2007, last retrieved 2007-04-03 from http://www.amiproject.org/ami-scientific-portal/documentation/annual-reports/pdf/SOTA-Localization-and-Tracking-Jan2007.pdf

4. Ahmad K, Taskaya-Temizel T, Cheng D, Gillam L, Ahmad S, et al. (2004) Financial Information Grid – an ESRC e-Social Science Pilot. In: Proceedings of the Third UK e-Science Programme All Hands Meeting (AHM 2004). Nottingham, United Kingdom.

5. Aono M, Sekiguchi Y, Yasuda Y, Suzuki N, Seki Y (2006) Time Series Data Mining for Multimodal Bio-Signal Data. IJCSNS International Journal of Computer Science and Network Security, Vol. 6 No. 10: 1-9

6. Baeza-Yates RA, Ribeiro-Neto BA (1999) Modern Information Retrieval. ACM Press / Addison-Wesley, New York

7. Blattner MM, Dannenberg RG (1990) CHI'90 Workshop on multimedia and multimodal interface design. SIGCHI Bulletin 22, 2: 54-58

8. Bounif H, Drutskyy O, Jouanot F, Spaccapietra S (2004) A multimodal database framework for multimedia meeting annotations. In: Multimedia Modelling Conference. IEEE Computer Society, Washington, DC, USA, pp 17- 25

9. CALO (2006) About CALO: SRI International, last modified 2006, last retrieved 2007-03-31 from http://caloproject.sri.com/about/

10. CHIL Project Description. Original Project Website, CHIL consortium, last retrieved 2007-06-25 from http://chil.server.de/servlet/is/104/

11. Cook P, Ellis C, Graf M, Rein G, Smith T (1987) Project Nick: meetings augmentation and analysis. ACM Transactions on Office Information Systems, Vo. 5, No. 2: 132-146

12. Costa CJ, Antunes PA, Dias JF (2001) A Model for Organisational Integration of Meeting Outcomes. In: Maung K. Sein, and others (eds) Contemporary Trends in Systems Development. Kluwer Plenum, 2001.

13. Fayyad U, Haussler D, Stolorz P (1996) Mining scientific data. Communication of the ACM Vol. 39, No. 11: 51-57

14. Fierrez-Aguilar J, Ortega-Garcia J, Garcia-Romero D, Gonzalez-Rodriguez J (2003) A Comparative Evaluation of Fusion Strategies for Multimodal Biometric Verification. Lecture Notes in Computer Science, Audio- and Video-Based Biometric Person Authentication, Volume 2688/2003. Springer, Berlin / Heidelberg, pp 830-837

15. García-Barrios VM, Gütl C (2006) Semantic Applications on MPEG-7 Descriptions of Multi-modal Meeting Corpora: First Results. Special Issue "Multimedia Metadata Community Workshop Results 2005 Dissemination", JUKM, Vo. 1, No. 1, pp. 45-53.

16. García-Barrios VM (2007) Personalisation in Adaptive E-Learning Systems. A Service-Oriented Solution Approach for Multi-Purpose User Modelling Systems.Ph.D. thesis, Graz University of Technology, Austria, May 2007.

17. Gütl C (2002) Approaches in modern knowledge discovery in the Internet. Ph.D. thesis, Graz University of Technology, Austria, September 2002.

18. Gütl C, García-Barrios VM (2005) Semantic Meeting Information Application: A Contribution for Enhanced Knowledge Transfer and Learning in Companies. In: Proceedings of 8th International Conference on Interactive Computer Aided Learning (ICL 2005), Villach, Austria

19. Gütl C, Safran C (2006) Personalized Access to Meeting Recordings for Knowledge Transfer and Learning Purposes in Companies. In: Proceedings of m-ICTE 2006.

20. Ho T, Antunes P (1999) Developing a Tool to Assist Electronic Facilitation of Decision-Making Groups. In: Proceedings of the String Processing and Information Retrieval Symposium & International Workshop on Groupware, pp 243 – 253

21. Jain R, Kim P, Li Z (2003) Experiential meeting system. In: Proceedings of the 2003 ACM SIGMM workshop on Experiential telepresence (ETP '03), pp. 1-12

22. Lalanne D, Lisowska A, Bruno E, Flynn M, Georgescul M, Guillemot M, and et.al. (2005) The IM2 Multimodal Meeting Browser Family. Technical report, March 2005, last retrieved 2007-04-03 from http://www.issco.unige.ch/projects/im2/mdm/docs/im2browsers-report-march2005.pdf

23. Lyman P, Hal RV (2003) How Much Information (Research Report). Last 2003-10-27, last retrieved 2007-06-07 from http://www.sims.berkeley.edu/how-much-info-2003

24. M4 Project. M4 Project Website, last retrieved 2007-06-23 from http://www.dcs.shef.ac.uk/spandh/projects/m4/overview.html

25. M4 (2005) Multimodal Meeting Manager. Deliverable D4.3: Report on Final Demonstrator and Evaluation, M4 Consortium, February 2005, last retrieved 2007-03-26 from http://www.dcs.shef.ac.uk/spandh/projects/m4/publicDelivs/D4-3.pdf

26. Magalhaes J, Pereira F (2004) Using MPEG standards for multimedia customization. Signal Processing: Image Communication. Vol. 19, no. 5: 437-456

27. McCobb G (2007) The W3C Multimodal Architecture, Part 1: Overview and challenges. Last edited 2007-05-08, last retrieved 2007-06-15 from http://www.ibm.com/developerworks/web/library/wa-multimodarch1/

28. Mader H (2007) Visualizing Multidimensional Metadata. Development of the Visualization Framework MD2VS. Master's thesis, Graz University of Technology, March 2007.

29. MISTRAL Official Website, MISTRAL research project, last retrieved 2007-07-07 from http://mistral-project.tugraz.at/

30. MMP (2003) Mapping Meetings project. Official Website, Columbia University, last modified 2003-08-11, last retrieved 2007-03-27 from http://labrosa.ee.columbia.edu/mapmeet/

31. Nigay L, Coutaz J (1993) A Design Space For Multimodal Systems: Concurrent Processing and Data Fusion. In: CHI '93: Proceedings of the SIGCHI conference on Human factors in computing systems, ACM Press, New York, NY, USA, pp 172-178

32. Nunamaker JF, Dennis AR, Valacich JS, Vogel DR, George JF (1991) Electronic meeting systems to support group work: theory and practice at Arizona. Communications of the ACM, Vo. 34, No. 7, 40-61

33. Nunamaker JF, Romano NC, Briggs RO (2001) A Framework for Collaboration and Knowledge Management. In: Proceedings of the 34th Hawaii International Conference on System Sciences.

34. Oviatt S (1999) Ten myths of multimodal interaction. Journal of the ACM, Vo. 42, No. 11: 74-81

35. Pallotta V, Ghorbel H, Ballim A, Lisowska A, Marchand-Maillet S (2004) Towards Meeting Informant Systems: Meeting Knowledge Management. In: International Conference on Enterprise Information Systems (ICEIS 2004)

36. Rienks R, Nijholt A, Barthelmess P (2007) Proactive Meeting Assistants: Attention Please! AI & Society. The Journal of Human-Centred Systems. Springer-Verlag. to appear
37. Romano NC, Nunamaker JF (2001) Meeting Analysis: Findings from Research and Practice. In: Proceedings of the 34th Annual Hawaii International Conference on System Sciences.
38. Rose RT, Quek F, Shi Y (2004) MacVisSTA: a system for multimodal analysis. In: ICMI '04: Proceedings of the 6th international conference on Multimodal interfaces, ACM Press, New York, NY, USA, pp 259-264
39. Sabol V, Granitzer M, Tochtermann K, Sarka W (2005) MISTRAL – Measurable, Intelligent and Reliable Semantic Extraction and Retrieval of Multimedia Data. In: 2nd European Workshop on the Integration of Knowledge, Semantic and Digital Media Technologies, London, UK, 2005
40. Sabol V, Gütl C, Neidhart N, Juffinger A, Klieber W, Granitzer M (2007) Visualization Metaphors for Multi-modal Meeting Data. In: Proceedings of the 6th Workshop of the Multimedia Metadata Community (WMSRM07), Aachen, Germany.
41. SAILLABS Speech Analytics: Maximizing the Value of Customer Communications. SAILLABS Technology, last retrieved 2007-07-15, http://www.sail-technology.com/index html?solutions/html/cm_sa.html
42. Srihari RK, Zhang Z, Rao A. (2000) Intelligent Indexing and Semantic Retrieval of Multimodal Documents. Information Retrieval, Vo. 2, No. 2-3: 245-275
43. Turoff M, Hiltz S (1977) Meeting through your computer. IEEE Spectrum, May 1977, 58-64
44. Wood C, Cross M, Phuong La (1998) Multimodal information retrieval, extraction and generation for usein the health domain. In: Knowledge-Based Intelligent Electronic Systems, 1998. Proceedings KES '98, Vo. 3, pp 307-316

GPSR Compliance
The European Union's (EU) General Product Safety Regulation (GPSR) is a set
of rules that requires consumer products to be safe and our obligations to
ensure this.

If you have any concerns about our products, you can contact us on

ProductSafety@springernature.com

In case Publisher is established outside the EU, the EU authorized
representative is:

Springer Nature Customer Service Center GmbH
Europaplatz 3
69115 Heidelberg, Germany